Connecting The Dots/Intelligent Trackers 2017 (CTD/WIT 2017)

AA004500

EPJ Web of Conferences Volume 150 (2017)

Orsay, France
6 - 9 March 2017

Editors:

Cecile Germain
Hadrien Grasland
Abdenour Lounis
David Rousseau
Dimitris Varouchas

ISBN: 978-1-5108-4641-8

Printed from e-media with permission by:

Curran Associates, Inc.
57 Morehouse Lane
Red Hook, NY 12571

Some format issues inherent in the e-media version may also appear in this print version.

Printed by Curran Associates, Inc. (2017)

For additional information, please contact EDP Sciences – Web of Conferences at the address below.

EDP Sciences – Web of Conferences
17, Avenue du Hoggar
Parc d'Activité de Courtabœuf
BP 112
F-91944 Les Ulis Cedex A
France

Phone: +33 (0) 1 69 18 75 75
Fax: +33 (0) 1 69 28 84 91

contact-edps@webofconferences.org

Additional copies of this publication are available from:

Curran Associates, Inc.
57 Morehouse Lane
Red Hook, NY 12571 USA
Phone: 845-758-0400
Fax: 845-758-2633
Email: curran@proceedings.com
Web: www.proceedings.com

TABLE OF CONTENTS

Preface

Cécile **Germain**[1,2], Hadrien **Grasland**[2], Abdenour **Lounis**[2], David **Rousseau**[2a] and Dimitris **Varouchas**[2]

[1] *LRI, Univ. Paris-Sud, CNRS & INRIA, Université Paris-Saclay, Orsay, France*
[2] *LAL, Univ. Paris-Sud, CNRS/IN2P3, Université Paris-Saclay, Orsay, France*

Abstract. Brief description of the workshop

1 Foreword

These are the conference proceedings of the workshop "Connecting The Dots/ Intelligent Trackers 2017" that was held at Orsay from March 6 to March 9, 2017. The web site with all transparencies remains available at :

http://ctdwit2017.lal.in2p3.fr.

With the parallel progress in pattern recognition algorithms and microelectronic technology, the design and performance of tracking detector is rooted in the solid interplay of hardware and software: sensors, readout and trigger electronics, online and offline reconstruction software. The main focus of the workshop was on pattern recognition and machine learning algorithms devoted to the reconstruction of particle tracks in high energy physics experiments, and the hardware developments that enable them.

This 2017 edition was a merger of the Connecting The Dot series (see CTD 2015 Berkeley[1] and CTD 2016 Vienna[2]) with the Workshop on Intelligent Tracker series (see WIT 2010 Berkeley[3], WIT 2012 Pisa[4] and WIT 2014 Penn[5]).

The workshop was plenary sessions only, with a mix of invited talks and accepted contributions. There were 100 participants from 16 countries. 46 talks were given and one hackathon on a simplified 2D tracking problem took place, yielding in total 16 proceedings, edited by the authors of this preface. Each proceeding was peer-reviewed by at least two referees, selected from the Program and Local committees.

Next workshop of Connecting The Dots series will take place at the University of Seattle (USA), early 2018.

---

[a] e-mail: rousseau@lal.in2p3.fr

[1] https://indico.physics.lbl.gov/indico/event/149/

[2] https://indico.hephy.oeaw.ac.at/event/86/

[3] https://indico.cern.ch/event/68677/

[4] https://indico.cern.ch/event/154525/

[5] https://indico.cern.ch/event/293354/

2 Program committee

The Program Committee members were: Alberto Annovi (INFN Pisa), Marina Artuso (Syracuse U.), Richard Brenner (Uppsala U.), David Brown (LBNL), Paolo Calafiura (LBNL), Michel De Cian (Heidelberg U.), Markus Elsing (CERN), Rüdolf Frühwirth (HEPHY Vienna), Frank Gaede (DESY), Maurice Garcia-Sciveres (LBNL), Martin Heck (KIT Karlsruhe), Benedict Hegner (CERN), Alessandro Marchioro (CERN), Fabrizio Palla (INFN Pisa), Eugenio Paoloni (INFN Pisa), Mark Pesaresi (Imperial College), Luciano Ristori (Fermilab), David Rousseau (LAL Orsay), Andre Schöning (Heidelberg U.) and Ariel Schwartzman (SLAC).

3 Local committee

The Local Committee members were: Valérie Brouillard (LAL Orsay), Cécile Germain (LRI & LAL Orsay), Hadrien Grasland (LAL Orsay), Isabelle Guyon (U. Paris Saclay), Abdenour Lounis (LAL Orsay), David Rousseau (LAL Orsay) and Dimitris Varouchas (LAL Orsay).

4 Sponsors

All members of all committees would like to thank for their support : the Faculté des Sciences d'Orsay Université Paris-Sud Université Paris-Saclay[6], the Paris-Saclay Center for Data Science [7], the LABoratoire d'EXcellence P2IO[8], l'INRIA[9], as well as, for its hospitality: the Laboratoire de l'Accélérateur Linéaire Univ. Paris-Sud, CNRS/IN2P3, Université Paris-Saclay, Orsay, France

[6]http://www.sciences.u-psud.fr/fr/index.html
[7]https://www.datascience-paris-saclay.fr
[8]http://www.labex-p2io.fr
[9]https://www.inria.fr

edp sciences

EPJ Web of Conferences

Statement of Peer review

In submitting conference proceedings to EPJ Web of Conferences, I certify to the Publisher that I adhere to the **Policy on Publishing Integrity** of the journal in order to safeguard good scientific practice in publishing.

1. All articles have been subjected to peer review administered by the proceedings editors.
2. Reviews have been conducted by expert referees, who have been requested to provide unbiased and constructive comments aimed, whenever possible, at improving the work.
3. Proceedings editors have taken all reasonable steps to ensure the quality of the materials they publish and their decision to accept or reject a paper for publication has been based only on the merits of the work and the relevance to the journal.

Title, date and place of the conference

Connecting The Dots / Intelligent Trackers, 6-9th March 2017, Orsay, France

Proceedings editor(s):

Cécile GERMAIN
Hadrien GRASLAND
Abdenour LOUNIS
David ROUSSEAU
Dimitris VAROUCHAS

Date and editor's signature

2017/07/11 DR

Expected Performance of Tracking in CMS at the HL-LHC

Erica **Brondolin**[1,a] on behalf of the CMS Collaboration

[1] *Institut fur Hochenergiephysik der ÖAW (AT), Vienna, Austria*

Abstract. During the Long Shutdown 3, scheduled from 2024 to mid 2026, CERN is planning an upgrade program in preparation of the High-Luminosity LHC (HL-LHC), which will bring the luminosity up to $5 \times 10^{34}\,cm^{-2}\,s^{-1}$, almost five times the one envisaged for 2017. As a consequence, up to 200 inelastic collisions on average will be superimposed on the event of interest. In this high-occupancy environment, reconstructing charged particle momenta with high precision is one of the biggest challenges. In order to face this new scenario, called Phase 2, the Compact Muon Solenoid (CMS) experiment will build a completely new silicon tracking detector. New approaches to track finding will have to be implemented to exploit the capabilities of the new tracker in addition to the algorithms already in use. The expected performance of CMS tracking at the HL-LHC is presented in this contribution.

1 Introduction

Starting in 2026, LHC will achieve a luminosity of $5 \times 10^{34}\,cm^{-2}\,s^{-1}$ with a bunch spacing of 25 ns. This LHC upgrade, called High-Luminosity LHC (HL-LHC), is accompanied by an upgrade program of the CMS experiment, called Phase 2, to maintain the excellent performance of the detector. In the specific case of tracking, the CMS tracker will be traversed by about 6000 charged particles with transverse momentum above 300 MeV at every cross beam, produced by about 200 collisions on average, that should be reconstructed. In order to face this challenge, CMS will install a new silicon-based tracker, which must satisfy the physics performance requirements under HL-LHC conditions.

2 The CMS tracking system in Phase 2

The present strip tracker was designed to operate efficiently up to an integrated luminosity of $500\,fb^{-1}$ and an average pile-up of 30 to 50 collisions per bunch crossing. The effect of the radiation damage and the resulting performance degradation is extensively described in Ref. [1]. The rise of the LHC luminosity will increase the average pile-up per bunch crossing to 140–200, and the integrated luminosity at the end of the HL-LHC era is estimated to be $3000\,fb^{-1}$. These completely new conditions will require a substantial upgrade of the CMS tracking system, which has to fulfill the following requirements:

[a] e-mail: erica.brondolin@cern.ch

- **High radiation tolerance**: The new tracker must be able to operate efficiently for the entire HL-LHC period up to an integrated luminosity of $3000\,\mathrm{fb}^{-1}$. While no maintenance interventions are foreseen for the Outer Tracker, the pixel modules of the Inner Tracker remain accessible and can be replaced.
- **Increased granularity**: This requirement is needed in order to avoid cluster merging and to maintain a good track separation in the Inner Tracker, and to limit the combinatorics in both the seeding and the track finding phase.
- **Contribution to the Level-1 trigger**: In Phase 2 the trigger system is expected to maintain an output rate of 100 kHz despite the tenfold luminosity. This appears to be possible only with the addition of tracking information to the CMS Level-1 trigger system.
- **Extended tracking acceptance**: As the CMS physics program will benefit from a larger acceptance region, the new tracker has been designed to cover the entire η region up to $|\eta| = 4$.
- **Reduced material**: The degradation of the tracking performance and the overall physics performance depends strongly on the amount of material in the tracking volume, therefore a lighter tracker option has been chosen. At $\eta = 0$ and in the most forward region, i.e. $3.0 < |\eta| < 4.0$, the total amount of material inside the tracking volume is estimated as 30% of a radiation length. In the Outer Tracker region, it rises up up to 80%.

Figure 1 shows one possible layout configuration of the Phase 2 tracker. It will be composed of an Inner Tracker based on silicon pixel modules, and of an Outer Tracker made from silicon modules with strip and macro-pixel sensors.

In this layout, the Inner Tracker has four barrel layers and eleven end-cap disks. Different pixel sizes in the Inner Tracker modules are under consideration, either $25\times100\,\mu\mathrm{m}^2$ or $50\times50\,\mu\mathrm{m}^2$, which will reduce the pixel area by a factor of about six compared to to the current CMS pixel detector. For the studies in this paper, the $25 \times 100\,\mu\mathrm{m}^2$ Inner Tracker modules have been simulated. The Outer Tracker has six barrel layers and five end-cap disks, where a part of the modules in the three innermost barrel layers are arranged in a tilted scheme. The Outer Tracker modules have been designed to provide useful information also at the first trigger level, allowing to maintain its acceptance rate below 100 kHz. Each module is a stack of two closely spaced sensors and is able to correlate hits produced in the two sensors in real time, thereby rejecting low-p_{T} particles using the strong bending power of the 3.8 T magnetic field. Two different kinds of modules for the final configuration of the Outer Tracker are under study: in the inner layers, macro pixel–strip (PS) modules composed of a bottom sensor with $1.5\,\mathrm{mm} \times 100\,\mu\mathrm{m}$ pixels and a top sensor with $2 \times 25\,\mathrm{mm}$ strips ($100\,\mu\mathrm{m}$ pitch); in the outer layers, strip–strip (2S) modules composed of two sensors with $2\times5\,\mathrm{cm}$ long strips ($90\,\mu\mathrm{m}$ pitch) each.

3 Track reconstruction method

The CMS track reconstruction algorithm is based on an iterative approach [2]. The main idea is to search for easier-to-find tracks first, to mask the hits associated to the found tracks, and to proceed to the next iteration. In this way the combinatorial problem is reduced, and the search for more difficult classes of tracks is simplified. Moreover, this approach introduces the possibility of developing special iterations that can improve track reconstruction in high-density environments such as jets, or to use the information from other subsystems such as muon chambers and calorimeters.

In each iteration, the so-called Combinatorial Track Finder is run. It can be divided into four different steps:

1. **Seed generation**: Using the information of three or four hits, the trajectory parameters and the corresponding uncertainties of the initial track candidates are computed.

2. **Track finding**: The current trajectory parameters are extrapolated to the next layer and possible compatible hits are found. Each of them is added to a clone of the track candidate. This step is repeated until more than one hit is missing or there are no more tracker layers.

3. **Track fitting**: The best possible estimates of the trajectory parameters at the *impact point*[1] is obtained using a final Kalman or Gaussian-sum filter plus the corresponding smoothers.

4. **Track selection**: Tracks are grouped in different classes (*Loose*, *Tight* or *High Purity*) according to various track quality cuts.

Although the track density is lower in the Outer Tracker, primary seeding in the Inner Tracker has the following advantages: even at very high pile-up, cluster merging is very much reduced thanks to the high granularity and the concomitant small channel occupancy; the Inner Tracker produces 3-D measurements that provide more constraints and better estimates of the initial trajectory parameters; the efficiency of the reconstruction of low momentum particles that do not cross the entire Outer Tracker is improved. The same seeding algorithm used in the present CMS tracking is also used for Phase 2: triplets are found by purely combining pixel hits in different layers, while quadruplets are created by the so-called "triplet propagation", which consists in propagating triplets to the next layer, finding possible compatible hits, computing the total χ^2 and rejecting hits above a certain χ^2 threshold. The seeding performance has been proved to have high efficiency with relatively low fake rate in realistic simulation studies.

The list of the tracking iterations can be found in Table 1. The main difference between iterations is the configuration of the seed generation and the target tracks. The seeding capability of the Outer Tracker is not yet exploited in the current Phase 2 track reconstruction; this topic is discussed in more detail in Section 5.

[1]The impact point is defined as the point of closest approach of the track to the assumed beam axis.

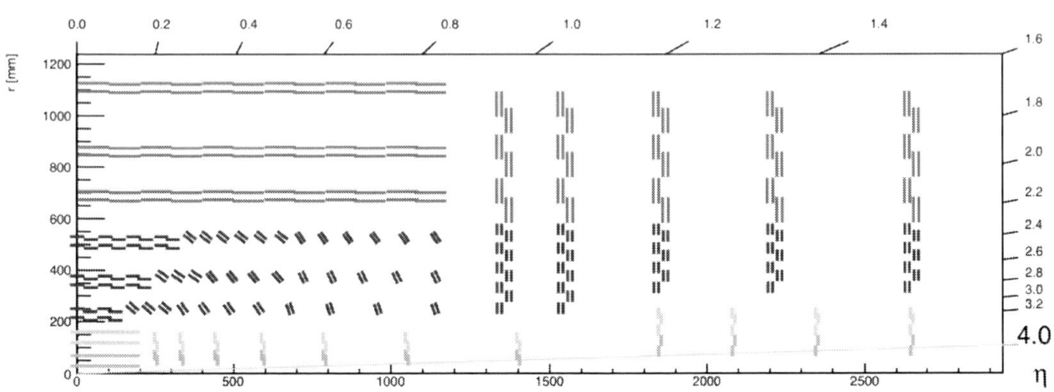

Figure 1. *R*-*z* view of a possible layout of the new CMS tracker foreseen for Phase 2. The inner part is composed of silicon pixel modules (yellow and light blue), while the outer part is composed by two different kind of silicon modules: PS modules (dark blue) and 2S modules (red).

Table 1. List of the six tracking iterations used in Phase 2, with the corresponding seeding configuration used and the target tracks.

Iteration	Step Name	Seeding	Target Track
0	HighPtQuadruplet	pixel quadruplets	prompt, high p_T
1	HighPtTriplet	pixel triplets	prompt, high p_T
2	DetachedQuadruplet	pixel quadruplets	displaced
3	LowPtQuadruplet	pixel quadruplets	prompt, low p_T
4	LowPtTriplet	pixel triplets	prompt, low p_T
5	Muon inside-out	muon-tagged tracks	muon tracks

4 Performance

4.1 Track reconstruction and parameters resolution

In order to assess the track reconstruction performance, defining the association between reconstructed and simulated tracks is particularly important. In the standard CMS definition, a reconstructed track is considered associated to a simulated particle if at least 75% of its hits originate from this simulated particle. If this is not the case, the reconstructed track is considered as a random combination of hits and marked as a fake track. The tracking efficiency is therefore defined as the fraction of charged particles associated to at least one reconstructed track, while the fake rate is the fraction of the fake tracks in the set of reconstructed tracks. Both efficiency and fake rate strongly depend not just on the algorithms and track selections used, but also on the intrinsic tracker properties, such as its layout, its material budget and the precision of its measurements.

The performance of the track reconstruction is evaluated using two different simulated samples. Where needed, a pile-up of minimum-bias events is superimposed, where the number of pile-up events is drawn from a Poisson distribution with mean equal to 140 or 200. The results are shown for the *High Purity* track selection and can be considered a conservative lower limit and are expected to improve with careful tuning of the reconstruction and matching algorithms for the high pile-up scenario.

The first sample considered consists of events with top quark pair production ($t\bar{t}$) and is used to assess the performance for a typical proton-proton collision at the HL-LHC. Figure 2 shows the results for this sample in both pile-up scenarios. In the top row on the left, the tracking efficiency for particles with $p_T > 0.9$ GeV is plotted as a function of the pseudorapidity, while on the right the rate of fake tracks with $p_T > 0.9$ GeV is shown. The efficiency is around 90% falling off at $|\eta| > 3.8$, while the fake rate is smaller than 5% in the entire range of η for pile-up 140. The small loss in efficiency around $\eta = 1.2$ is partially due to the pixel configuration and will be fully recovered by extending the seeding to the Outer Tracker. In the bottom row efficiency and fake rate are plotted as a function of the transverse momentum. The tracking efficiency is around 90% in the range between 1 GeV and 100 GeV, and the fake rate is smaller than 10% below 10 GeV with 140 pile-up events on average. As most particles in the $t\bar{t}$ events are hadrons, the deterioration of the efficiency at high p_T can be explained by the production of secondary particles in inelastic processes, which are very close to the mother particle at high energies and mislead the pattern recognition. With higher pile-up, the efficiency is not affected, while the fake rate almost doubles, but is still tolerable. Figure 3 shows the contribution of the individual iterations to the final tracking efficiency of $t\bar{t}$ events with pile-up 140. Thanks to the iterations targeting low transverse momentum particles, the efficiency is at least 30% above 300 MeV/*c*.

The second sample consists of single muons, generated according to a flat η distribution in the range $|\eta| < 4.0$, with fixed transverse momenta of 10 GeV. It is used to test the tracking performance

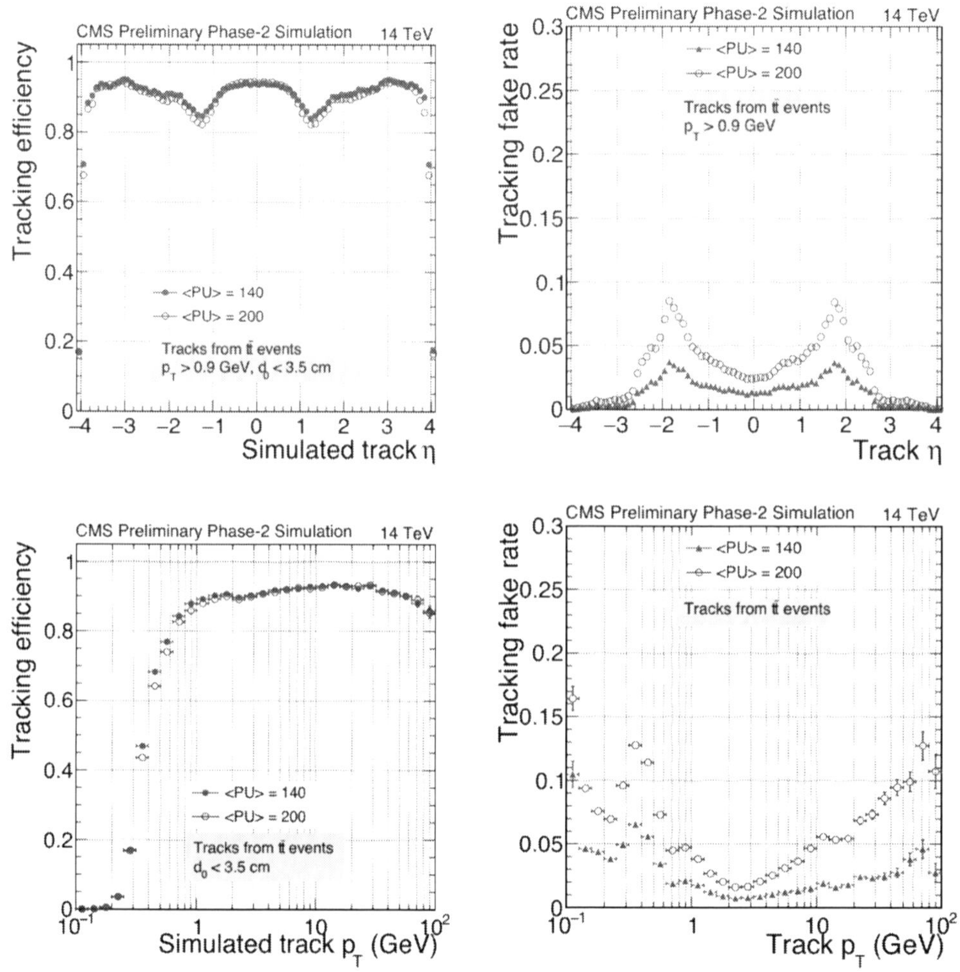

Figure 2. Tracking efficiency (left) and fake rate (right) as a function of the pseudorapidity (upper row) and as a function of the transverse momentum (lower row) for $t\bar{t}$ events simulated for the Phase 2 detector with superimposed pile-up collisions of 140 (full circles) and 200 (open circles). Tracks passing the *High Purity* quality requirements and produced less than 3.5 cm from the centre of the beam spot in radial direction are used. The performance as a function of η is for generated particles with $p_T > 0.9$ GeV.

as a function of the pseudorapidity and to determine the resolution of the estimated track parameters at the impact point. In the CMS convention the five trajectory parameters are: d_0, the transverse impact parameter; z_0, the longitudinal impact parameter; ϕ, the track angle in the transverse plane; $\cot\theta$, the cotangent of the polar angle; and p_T, the transverse momentum. The resolution of track parameters is given by the RMS of the residuals, i.e. the difference between the estimated and the simulated track parameters. Figure 4 shows the tracking efficiency for single muons with a transverse momentum of 10 GeV in both pile-up scenarios. It can be seen that the single muon efficiency is independent of the amount of pile-up. Figure 5 shows the resolution of the five parameters for single muons with $p_T = 10$ GeV as a function of the pseudorapidity for both the current tracker and the future Phase 2

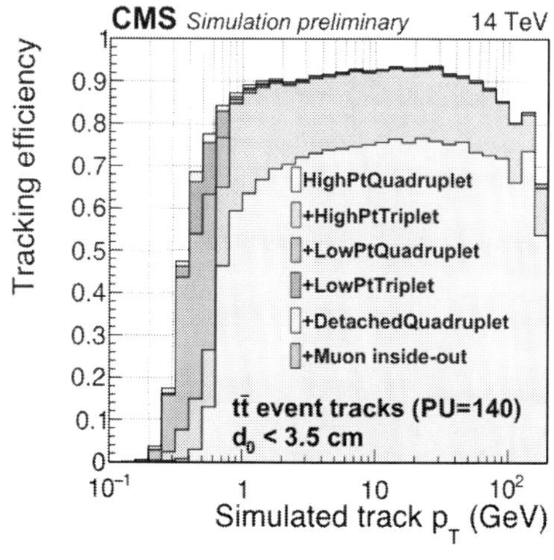

Figure 3. Tracking efficiency of individual iterations for the $t\bar{t}$ sample with an average pile-up of 140, reconstructed with the Phase 2 detector. The label of each iteration identifies the seeding algorithm used (see Table 1).

tracker. The resolution deteriorates for large η for all parameters, but for different reasons: the shorter lever arm for the estimation of the curvature and of the azimutal angle ϕ; the larger extrapolation distance from the innermost hit to the beam axis for the impact parameter; and the parametrization of the polar angle θ by $\cot\theta$. In all five parameters, the resolution is considerably better with the Phase 2 tracker. In particular, a significantly improved p_T resolution can be achieved thanks to the reduction of the material budget and a better hit resolution. Also the transverse impact parameter resolution is much better than with the Phase 1 detector, ranging from below 10 μm in the barrel region to about 20 μm in the very forward region.

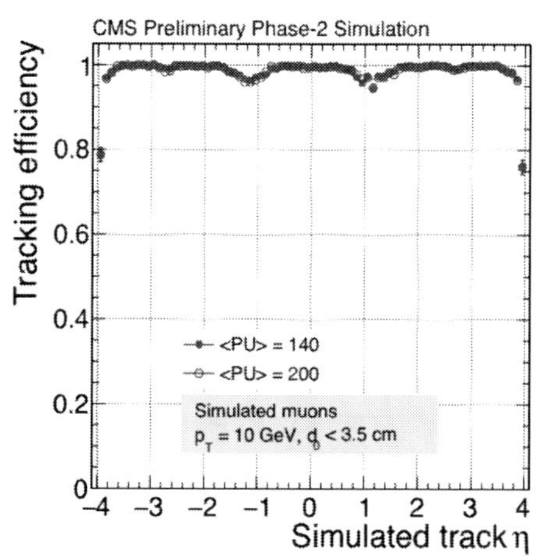

Figure 4. Tracking efficiency as a function of the pseudorapidity for single muons with p_T equal to 10 GeV and pile-up of 140 (full circles) and 200 (open circles).

4.2 Signal vertex reconstruction and resolution

The Phase 2 vertex reconstruction [2] is based on the current algorithm used in the present CMS tracker. It proceeds in two steps: first, the vertices are found as clusters of tracks coming from the same interaction point using the Deterministic Annealing algorithm [3]; then, for each found vertex, an Adaptive Vertex Fit [4] computes the best estimate of the vertex position and of the parameters of the associated tracks without using any information on the beam position. Finally, the reconstructed vertices are sorted to identify the *signal vertex* among the pile-up vertices. The vertex sorting is based on the $\sum p_{\mathrm{T}}^2$. In the case of jets, the individual tracks contributing to the $\sum p_{\mathrm{T}}^2$ are replaced by jets obtained by clustering tracks originating from the same vertex. The final $\sum p_{\mathrm{T}}^2$ is computed using these jets, the remaining isolated single tracks and the missing transverse momentum, re-weighed accordingly. A more detailed description of the sorting method can be found in Ref. [1].

The probability of reconstructing the signal vertex and to tag it correctly is an important figure of merit, and it is strongly correlated with the amount of pile-up in the event. For $t\bar{t}$ events, the efficiency of the correct signal vertex identification is 94% with 140 pile-up events, and 89% with 200 pile-up events. The quality of the vertex reconstruction is also shown in Figure 6. On the left, the number of reconstructed vertices as a function of the number of simulated vertices is shown. The dependence is almost linear. On the right, the merging rate, i.e. the probability of two adjacent vertices to merge into a single one, is shown as a function of their distance in the z coordinate. The merging rate drops to zero at about 0.6 mm both with 140 and with 200 pile-up events. Note that the merging rate is smaller than one even at distances below 25 μm.

Another important aspect is the resolution of the vertex position in the x, y and z coordinates, which is shown in Figure 7 as a function of the number of tracks associated to the vertex. It can be seen that the resolution of the vertex position is almost independent of the amount of pile-up in the event. Moreover, the longitudinal resolution is only 50% worse than the transverse one, as expected given the pixel dimensions of Inner Tracker modules.

As in the case of the tracking performance, all results should be considered as a lower limit of the expected performance.

5 Future developments

Here some developments are listed that are envisaged for further improvement of the CMS Phase 2 tracking:

- **Tracking@HLT using GPUs**: The software for the CMS High-Level Trigger (HLT) tracking has been moved in 2017 from a one-event-per-thread strategy to a many-threads-per-event approach. This massive parallelism within each event scales with the pile-up by offloading the combinatorics to as many threads as are available. This task has been implemented in CMS designing a high-level trigger based on GPUs. The same valid approach will be used also for the future Phase 2 HLT.
- **Tracking in the Outer Tracker**: Many developments are made possible by this new kind of detector. For example, the L1 track trigger information can be used directly in the offline reconstruction for instance to create seeds from L1 tracks or to incorporate L1 algorithms in the offline tracking. Also, the modules of the new Outer Tracker can be used to build short track segments (*vector hits*) reconstructed from compatible hits in the two sensors of a module, which contain not only position information but also direction information. Algorithms are under development that exploit this information for seeding in the Outer Tracker and for searching for displaced tracks without using pixel information.

- **Multiple algorithms**: This exploits the concept of using different algorithms specialized for certain tasks in different iterations.
- **Iterations targeting physics objects**: As already mentioned in Section 3, the iterative approach is very suitable for introducing special iterations targeting specific physics object. This has already been tested in the current CMS tracking, but has not yet been fully introduced into the Phase 2 reconstruction sequence. Some examples are the muon specific iterations and the reconstruction of tracks inside jet cones, which makes identification of physics objects more robust, and significantly increases the track reconstruction efficiency and purity [5].

References

[1] D. Contardo, M. Klute, J. Mans, L. Silvestris, J. Butler (CMS), CERN-LHCC-2015-010, LHCC-P-008, CMS-TDR-15-02 (2015)

[2] The CMS Collaboration (CMS), JINST **9**, P10009 (2014)

[3] K. Rose, Proceedings of the IEEE **86/11**, 2210 (1998)

[4] R. Frühwirth, W. Waltenberger, P. Vanlaer, Tech. Rep. CMS-NOTE-2007-008, CERN, Geneva (2007), `https://cds.cern.ch/record/1027031`

[5] The CMS collaboration, https://twiki.cern.ch/twiki/bin/view/CMSPublic/HighPtTrackingDP

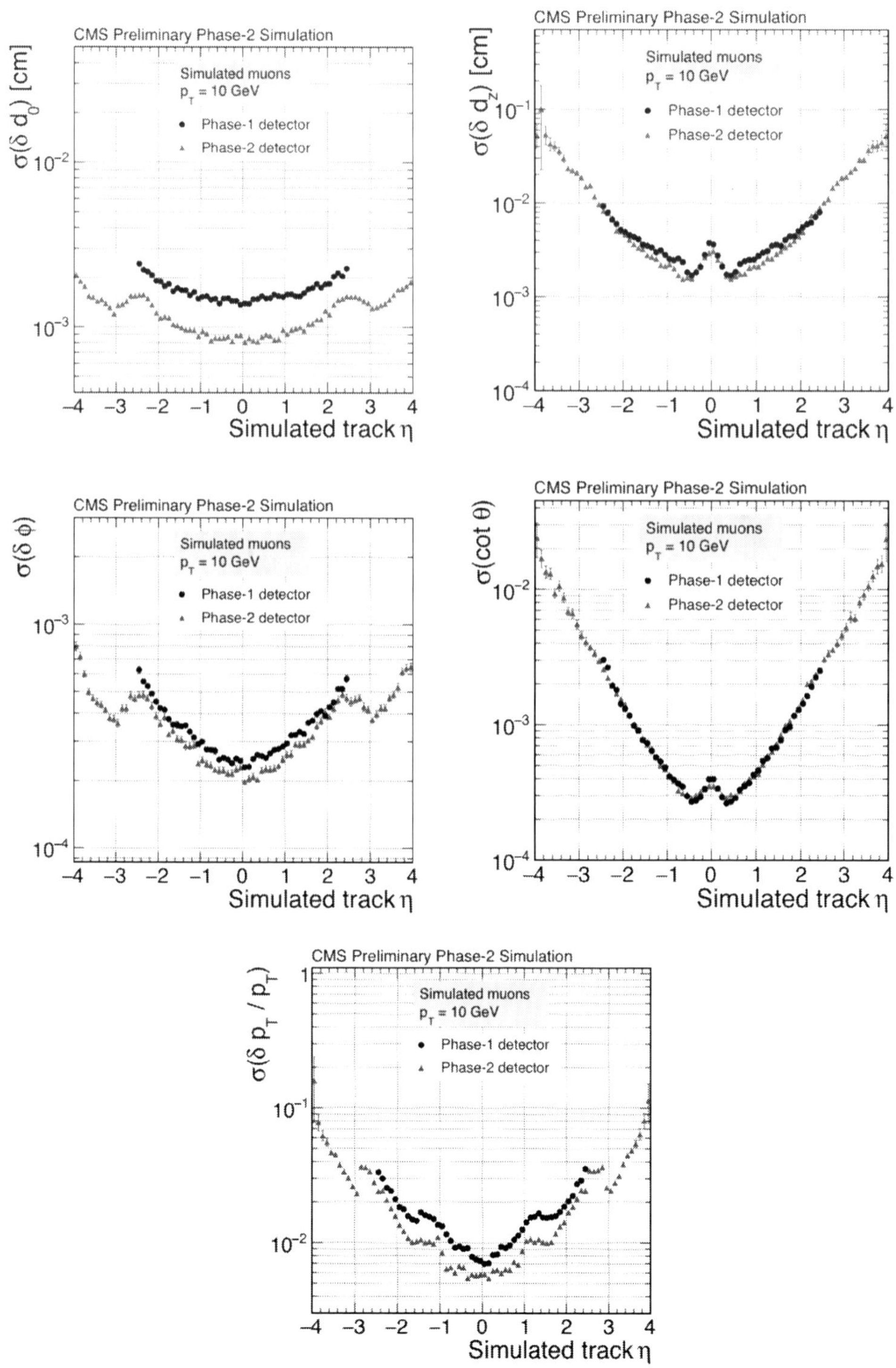

Figure 5. Comparison of the resolution, as a function of pseudorapidity, of the five track parameters for the current (black) and the upgraded (red) tracker for single, isolated muons with transverse momenta of 10 GeV. From top to bottom and left to right:transverse and longitudinal impact parameters, ϕ, $\cot\theta$ and p_T.

 DOI: 10.1051/epjconf/201715000001

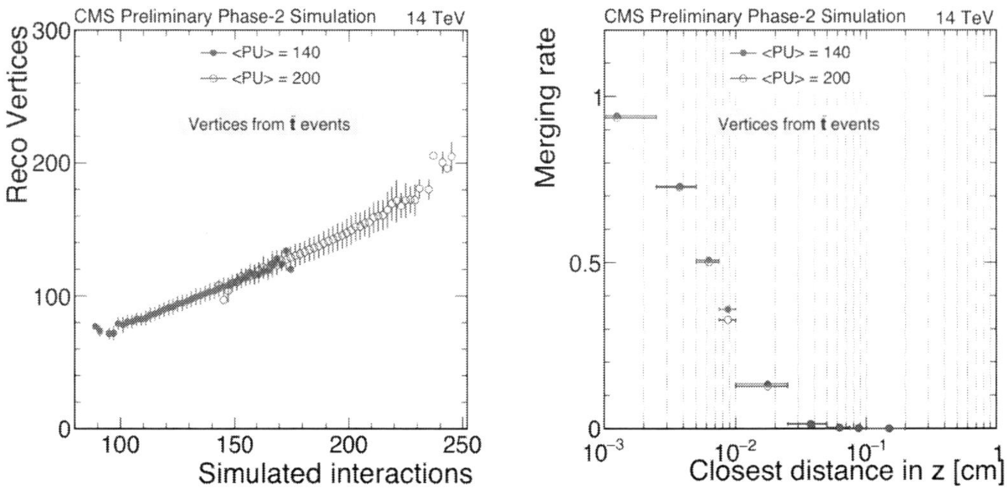

Figure 6. Number of reconstructed ("Reco") vertices versus the number of simulated vertices (left) and merging rate of vertices as a function of their distance in the z coordinate (right) for $t\bar{t}$ events with 140 superimposed pile-up events (full circles) and 200 pile-up events (open circles).

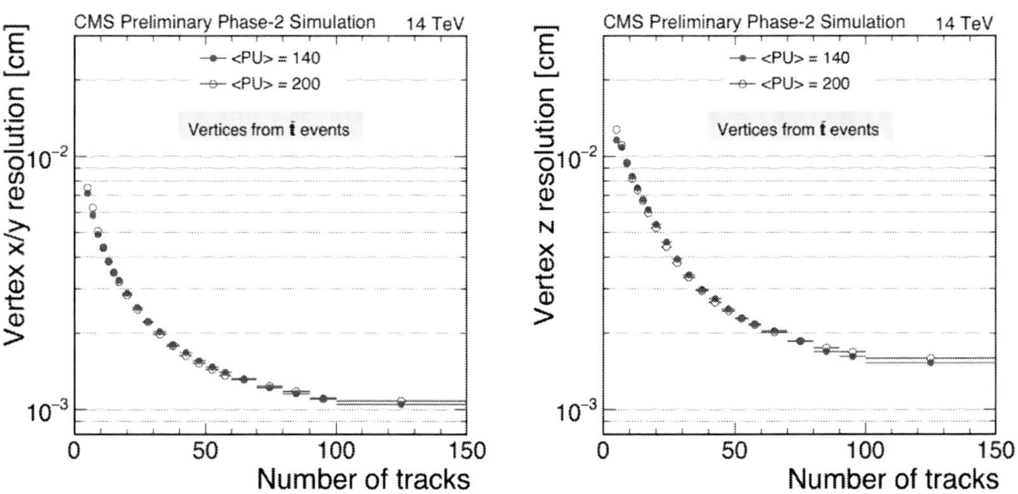

Figure 7. Vertex position resolution in x and y (left) and z (right) as a function of the number of tracks associated to the vertex, for $t\bar{t}$ events with 140 pile-up events (full circles) and 200 pile-up events (open circles).

Wireless data transmission for high energy physics applications

Sebastian **Dittmeier**[1,a], Richard **Brenner**[2], Dragos **Dancila**[2], Cedric **Dehos**[3], Patrick **De Lurgio**[4], Zelimir **Djurcic**[4], Gary **Drake**[4], Jose Luis **Gonzalez Gimenez**[3], Leif **Gustafsson**[2], Do-Won **Kim**[5], Elizabeth **Locci**[6], Ullrich **Pfeiffer**[7], Dieter **Röhrich**[8], Anders **Rydberg**[2], André **Schöning**[1], Alexandre **Siligaris**[3], Hans Kristian **Soltveit**[1], Kjetil **Ullaland**[8], Pierre **Vincent**[3], Pedro **Rodriguez Vazquez**[7], Dirk **Wiedner**[1], and Shiming **Yang**[7]

[1] *Physics Institute, Heidelberg University, Germany*
[2] *Department of Physics and Astronomy, Uppsala University, Sweden*
[3] *CEA/LETI/DRT/DACLE/LAIR, Grenoble, France*
[4] *Argonne National Laboratory, Argonne, IL 60439, USA*
[5] *Gangneung National University, Korea*
[6] *CEA/DRF/IRFU/SPP, Gif-sur-Yvette, France*
[7] *Institute for High-Frequency and Communication Technology, University of Wuppertal, Germany*
[8] *Department of Physics and Technology, University of Bergen, Norway*

Abstract. Silicon tracking detectors operated at high luminosity collider experiments pose a challenge for current and future readout systems regarding bandwidth, radiation, space and power constraints. With the latest developments in wireless communications, wireless readout systems might be an attractive alternative to commonly used wired optical and copper based readout architectures.

The WADAPT group (Wireless Allowing Data and Power Transmission) has been formed to study the feasibility of wireless data transmission for future tracking detectors. These proceedings cover current developments focused on communication in the 60 GHz band. This frequency band offers a high bandwidth, a small form factor and an already mature technology. Motivation for wireless data transmission for high energy physics application and the developments towards a demonstrator prototype are summarized. Feasibility studies concerning the construction and operation of a wireless transceiver system have been performed. Data transmission tests with a transceiver prototype operating at even higher frequencies in the 240 GHz band are described. Data transmission at rates up to 10 Gb/s have been obtained successfully using binary phase shift keying.

[a]e-mail: dittmeier@physi.uni-heidelberg.de

1 Introduction

High Energy Physics (HEP) experiments, as they are conducted at the LHC[1], the HL-LHC[2] or at potential future colliders, are operated at ever increasing luminosities and event rates. Correspondingly, more and more data are generated by these experiments that have to be read out and processed. Especially highly granular silicon detectors as they are used for vertexing and tracking pose a challenge for readout systems.

Today, readout systems typically use wired optical data transmission for data rates beyond 1 Gb/s per channel. However, radiation hardness issues, space and power constraints limit the usage of optical components in very dense pixel detector environments to places far outside the active volume. This is the case, for instance, for the vertex detector for the Phase-II upgrade of the ATLAS Inner Tracker (ITk) [1]. In these cases, wired copper transmission lines are used to transfer the data over several metres towards the optical drivers. Aiming for very high data rates, copper based transmission lines typically suffer from an increased impedance which limits the transmission of data rates of a few Gb/s to distances of a few metres. Developments in both wired optical and copper technologies are ongoing, but a third, alternative technology relying on wireless data transmission might be attractive for future high rate readout systems in dense and hostile environments, as first described in [2]. Regarding bandwidth, wireless technologies have caught up and can deliver rates in the range of several gigabits per second, as well.

Wireless data transmission could help to reduce the number of cables, optical or copper, used for data transmission as well as their connectors, which are often prone to damage. With less cables and connectors, installation and repair of such a complex system like a tracking detector can be simplified. Moreover, cables contribute significantly to the material budget in the active detector volume, which leads to multiple scattering and acts as a source of secondary particles.

Wireless data transmission would further allow to realize completely different readout topologies [2]. To keep a distinct level of modularity in a layer based silicon tracking detector, different layers can not be connected by wires within the active volume. Thus, directly in the active volume

[1]Large Hadron Collider
[2]High Luminosity-Large Hadron Collider

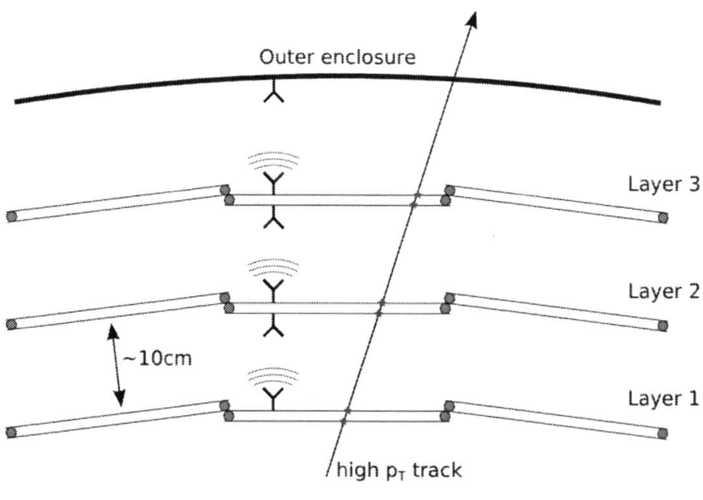

Figure 1. Conceptual sketch of a wireless radial readout of a cylindrical tracking detector, adapted from [2].

no communication between layers is possible and data has to be transferred to the outside, first. A wireless readout system would enable layers to communicate with each other. This could be used to enhance on-detector intelligence such as finding of tracks with high transverse momentum online as the data could follow the particle trajectories as depicted in figure 1. Fast track trigger applications might benefit from this readout topology [2].

Broadcasts of signals that are common in the detector could also be easily distributed using wireless data transmission, as a signal emitted by a single transmitter can be received by several receivers at the same time. This could be interesting for detector synchronization or to distribute a system-wide reset or trigger signal.

These motivations have led to the formation of the WADAPT (Wireless Allowing Data and Power Transmission) consortium. The goal of this research group is to identify specific needs of projects that might benefit from wireless technologies with the objective of providing a common platform for research and development [3]. Along that line, a wireless data transmission demonstrator for HEP applications is developed and shortly described in section 3. Studies to demonstrate the feasibility of this technology for HEP are summarized in section 4. Section 5 shows some further prospects of wireless technologies operating at carrier frequencies above 200 GHz.

2 Wireless data transmission

Over the last years, wireless data transmission technologies have seen tremendous progress to handle the ever increasing demand on bandwidth. The latest developments in wireless standardization by IEEE foresee data rates beyond one gigabit per second [4], and this trend of increasing wireless data rate capabilities seems to continue even further [5].

The Shannon-Hartley theorem gives a limit on the maximum data rate of a transmission channel. It depends on two factors: the bandwidth, and the average signal to noise ratio over that bandwidth. For typical Wi-Fi at home, operating at carrier frequencies of 2.4 GHz or 5 GHz, the bandwidth available per wireless channel is limited to a few tens of MHz. To achieve high data rates modulation schemes with high spectral efficiency[3] have been developed. A reliable data transmission with high spectral efficiency, however, requires complex signal processing and a high signal-to-noise ratio. Both typically come at the cost of a significant power consumption.

Improvements in silicon technologies over the last decades allow to produce ASICs operating at higher radio frequencies (RF). This also allows to allocate a larger bandwidth for a wireless channel. Especially the mm-wave band is of great interest to industry and appears also very attractive to the HEP community. In the 60 GHz band a total bandwidth of 9 GHz is available. This allows for high data rates even with simple modulation schemes like On-Off-Keying (OOK). Thus, RF devices can be produced without complex signal processing cicuitry and therefore low power consumption. The high frequencies furthermore allow for a high degree of miniaturization. For instance, antennas can be directly integrated in the chip. The free space path loss is significantly larger than at 2.4 GHz, thus the 60 GHz band is very well suited for safe, short distance communication.

Various examples of 60 GHz transceivers can be found in literature [6–9]. However, in order to qualify for HEP applications, a specific transceiver is required which fulfils the stringent constraints on power consumption and should be able to operate in extremely hostile radiation environments. For this application, a demonstrator prototype is currently under development.

[3]Spectral efficiency is defined as the number of bits that can be transferred per second and unit of bandwidth.

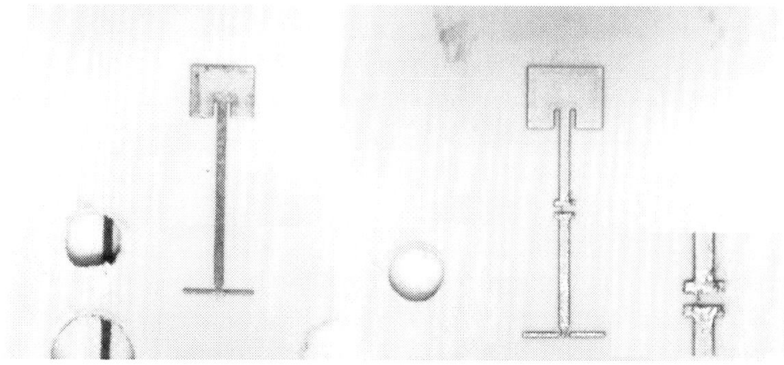

Figure 2. The two antennas under test for the bond wire study [10]. The left antenna has a solid transmission line, while the transmission line of the right antenna is broken intentionally by a gap with a compensation circuit. In the lower right corner, a zoomed-in view of the gap bridged by a single bonding wire can be seen.

3 A 60 GHz demonstrator for HEP

The purpose of the demonstrator is to validate the feasibility of wireless data transmission for HEP applications. It has to fulfil several requirements: It must be operational after irradiation up to very high doses as expected for the HL-LHC. It has to provide data rates of the order of 5 Gb/s, similar to what is expected per link from the silicon detectors of the large experiments at HL-LHC. At the same time power efficiency is crucial. In order to benefit from wireless data transmission, the power consumption of the on-detector readout part should be reduced compared to commonly used readout architectures.

At Heidelberg University a wireless transceiver chip is under development [11, 12]. The prototype will be produced in the IBM SiGe BiCMOS 130 nm 8HP process. Utilizing the full 9 GHz bandwidth and On-Off-Keying (OOK) modulation, the target data rate of this device is 4.5 Gb/s. The targeted power consumption is about 150 mW for a data transmission over distances between 20 cm and 1 m. This corresponds to an energy efficiency of about 33 pJ/bit. The actual transmission distance that can be achieved for a given power consumption, however, is largely influenced by the choice of the external antenna.

For the demonstrator, a printed circuit board will be designed with an on-board antenna. Directive patch antennas have been successfully fabricated and simulations of the reflection loss have been verified in measurements [13]. The connection between the RF output of the transceiver and the antenna on the demonstrator will potentially be realized using wire bonds. Therefore, a wire bonding study is performed using the patch antennas shown in figure 2. The impedance of a standard wire bond is found not to match the antenna under test properly, but signal losses can be compensated using an additional second wire and a compensation circuit, see figure 3 [10].

4 Studies in the 60 GHz band

Several studies have been conducted with the scope of testing the feasibility of wireless data transmission in the 60 GHz band for HEP applications. These include simple data transmission studies as well as interference and link density studies.

High speed data transmission was tested with two setups. Using a commercially available transmitter and receiver chipset by Hittite, an error-free 1.76 Gb/s link was established at a transmission

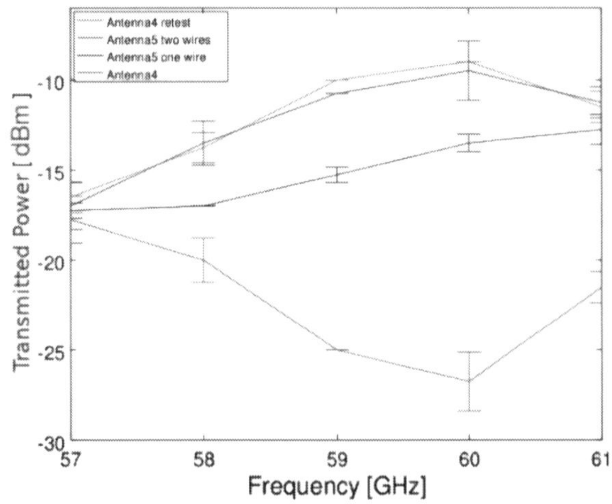

Figure 3. Measured intensity for different antenna configurations [10]. Purple curve: Antenna with solid transmission line (figure 2, left) serves as reference. Black curve: Antenna with gap included in the transmission line (figure 2, right image) and the gap is bridged by a single wire bond. Blue curve: The same antenna with gap, but with two wires to close the gap. The full signal is recovered by using two bond wires instead of one. The red curve belongs to the same antenna as the purple curve, but part of the signal was lost due to an issue with the etching process.

distance of 22 cm. An upper limit on the bit error rate was measured to be at the 10^{-14} level [14]. CEA-LETI demonstrated data transmission at a rate of 3.8 Gb/s using a transceiver produced in 65 nm CMOS technology [6].

A wireless detector readout scheme as illustrated in figure 1 relies on the simultaneous use of the same frequencies between different layers to fully exploit the available bandwidth. This is possible as typical silicon detector modules are metallized and thus fully reflective for waves in the 60 GHz band. This can be confirmed by a measurement of the transmission loss through an ATLAS SCT module, which is found to be at least 50 dB in the frequency range from 57.3 to 61.3 GHz, see figure 4 [14].

While enabling to re-use frequencies between different layers, the high reflectivity of the modules can potentially lead to cross talk between neighbouring links on the same layer. But the free space path loss in the 60 GHz band is already so large that only links that are very close can potentially interfere. In addition, the detector environment has one clear advantage compared to mobile applications: Once the system is installed, it is completely stationary. This allows to perform design studies and ray-tracing simulations with detailed detector models beforehand to reduce interferences and therefore optimize the overall bandwidth usage.

There are several means that can be utilized for the system design when aiming at very high channel densities. The emitted power of the transmitters can be adjusted such that the receivers' signal margin is only met by intended links and not by cross talk. Furthermore, highly directional antennas and RF signal polarization can be used to reduce interferences. Absorbing materials like low-density graphite foam can be used to attenuate reflections. A study was performed demonstrating that, by applying these means, two wireless links can be safely operated in parallel at a pitch of 5 cm between two highly reflective layers that are 10 cm apart, resembling a typical tracking detector scenario. A more detailed description of these studies can be found in [14].

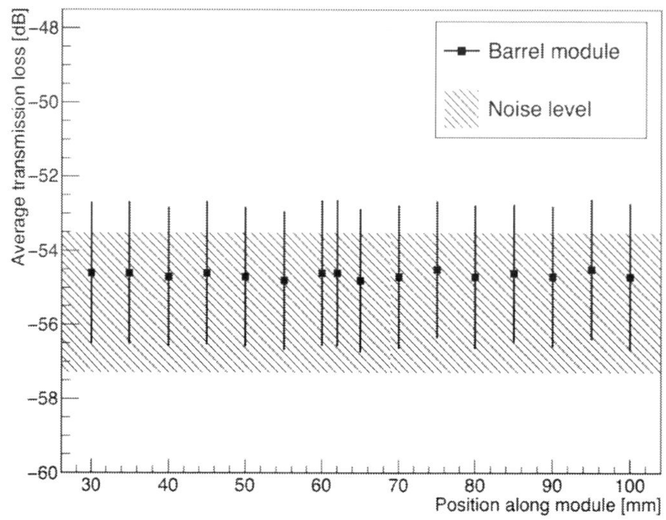

Figure 4. Transmission loss of mm-waves through an ATLAS SCT barrel module. The transmission loss is averaged over the frequency range from 57.3 to 61.3 GHz and measured over different positions along the module. The noise level indicates the minimum sensitivity of the spectrum analyser. The uncertainties represent the RMS of the averaged measurements. Plot taken from [14].

In order to further qualify the 60 GHz technology for HEP applications, it has to be ensured that the wireless links do not disturb the detector's electronics. Tests with an ATLAS ITK hybrid with ABCn130 readout ASICs showed no observable effect on the module's performance [14]. A similar measurement was conducted with an HV-MAPS prototype developed for Mu3e. No effect on this sensor's performance due to the wireless link was observed, as well [14]. We therefore conclude that wireless links at 60 GHz are a safe readout option for state of the art silicon detectors.

5 Prospects of carrier frequencies above 200 GHz

Recent developments in silicon technologies allow for even higher carrier frequencies, which, in turn, allow to use an even larger bandwidth. Examples of wireless transceivers operating at 210 to 283 GHz can be found in [15–17].

We performed data transmission studies with a 240 GHz transmitter and receiver chipset [16]. The chipsets are glued and wire bonded to printed circuit boards. The chipsets feature an integrated ring antenna for signal transmission and reception. To increase the antenna's directivity, silicon lenses with a gain of more than 25 dBi are mounted just below the circuits. A transmission distance of about 40 cm was realized. A photo of the setup in the laboratory is shown in figure 5. The intermediate frequency (IF) bandwidth was characterized at a fixed carrier frequency of 235 GHz using an intermediate frequency sweep in the range of 0 to 20 GHz and measuring the received signal in a spectrum analyser. The 3/6-dB IF bandwidth was measured to be 12/13 GHz, respectively, see figure 6.

This large bandwidth allows for data transfer at very high rates without the use of a complex modulation scheme. We performed bit error rate tests of a pseudo random bit pattern (PRBS-7) using binary phase shift keying (BPSK). Up to 8 Gb/s not a single bit error was observed, giving an upper

Figure 5. Wireless link using the chipsets described in [16] at a transmission distance of about 40 cm in the laboratory.

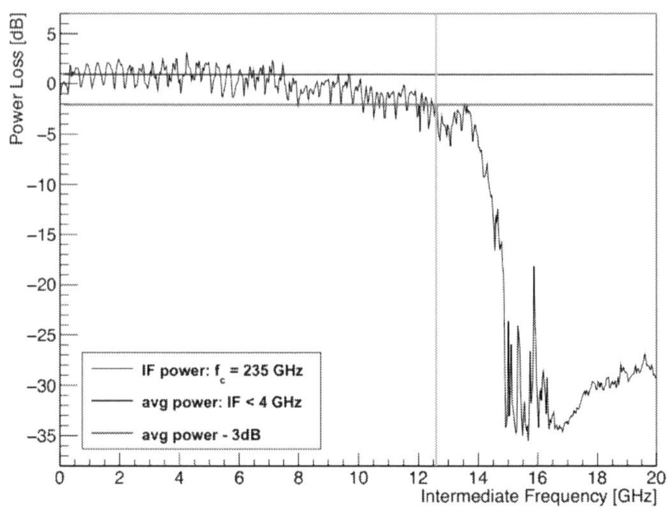

Figure 6. The intermediate frequency spectrum measured with a spectrum analyser at the 240 GHz receiver. The intermediate frequency of the transmitter is varied between 0 to 20 GHz. The green line highlights the 3 dB bandwidth.

limit on the bit error rate of about 10^{-13}. At 10 Gb/s a bit error rate of 10^{-14} was measured. These measurements demonstrate the enormous potential of wireless data transmission at high frequencies.

However, the technology of transceivers operating above 200 GHz is still in the prototyping phase. To be fully competitive, the energy efficiency has to be improved. The link tested here consumes about 2.8 W, corresponding to 280 pJ/bit.

6 Summary and outlook

There has been huge progress in wireless data transmission technologies over the last couple of years. The mm-waves developments allow for extremely high data rates while also being very power efficient.

The advantages of a wireless readout system compared to wired ones as used in HEP today are manifold. The reduction of cables and connectors would help to reduce dead material inside detectors as well as ease the detector's installation. New readout topologies for tracking detectors are possible which have great potential of enhancing fast track trigger capabilities.

A 60 GHz demonstrator is currently developed to show the potential of wireless data transmission for HEP applications. Studies to test the feasibility of wireless data transfer for tracking detectors have been conducted. High speed data transmission has been demonstrated with two setups at data rates of 1.76 Gb/s and 3.8 Gb/s. Carrier frequencies above 200 GHz allow for even higher data rates. A transmission at a data rate of 10 Gb/s has been demonstrated. For high density link environments in general, signal integrity can be ensured by exploiting directional antennas, polarization or attenuating reflections using an absorbing foam. In order to apply wireless data transfer in future detectors, detailed design studies will have to be conducted to demonstrate the full potential of wireless communication and to create a working wireless readout system.

References

[1] M. Wensing (ATLAS), JINST **12**, C01093 (2017)
[2] R. Brenner, S. Cheng, JINST **5**, C07002 (2010)
[3] R. Brenner et al. (WADAPT) (2015), `1511.05807`
[4] IEEE Computer Society, doi:10.1109/IEEESTD.2012.6392842 (2012)
[5] ISSCC 2016 Trends (2016), `http://isscc.org/doc/2016/ISSCC2016_TechTrends.pdf`
[6] A. Siligaris et al., IEEE Journal of Solid-State Circuits (JSSC) (2011)
[7] C. W. Byeon, C. H. Yoon, C.S. Park, IEEE Trans. Microw. Theory Techn. **61** (2013)
[8] A. Tomkins et al., Custom Integrated Circuits Conf. (CICC) (2008)
[9] C. Marcu et al., Proceedings of International Solid State Circuits Conference - ISSCC 2009 (2009)
[10] R. Brenner (2017), `https://indico.cern.ch/event/608206/contributions/2451744/attachments/1402081/2140369/Wifi_meeting20170127_wirebond.pdf`
[11] H.K. Soltveit, R. Brenner, A. Schöning, D. Wiedner, JINST **7**, C12016 (2012)
[12] H.K. Soltveit, S. Dittmeier, A. Schöning, D. Wiedner, Nuclear Science Symposium and Medical Imaging Conference (NSS/MIC), 2013 IEEE pp. 1–6 (2013)
[13] D. Pelikan, N. Bingefors, R. Brenner, D. Dancila, L. Gustafsson, JINST **9**, C11008 (2014)
[14] S. Dittmeier, A. Schöning, H.K. Soltveit, D. Wiedner, Nucl. Instrum. Meth. **A830**, 417 (2016), `1604.06259`
[15] Z. Wang, P.Y. Chiang, P. Nazari, C.C. Wang, Z. Chen, P. Heydari, IEEE Journal of Solid-State Circuits **49**, 564 (2014)
[16] N. Sarmah, P.R. Vazquez, J. Grzyb, W. Foerster, B. Heinemann, U.R. Pfeiffer, 2016 11th European Microwave Integrated Circuits Conference (EuMIC) (2016)
[17] J.M. Guerra, A. Siligaris, J.F. Lampin, F. Danneville, P. Vincent, 2013 IEEE Radio Frequency Integrated Circuits Symposium (RFIC) pp. 301–304 (2013)

The HEP.TrkX Project: deep neural networks for HL-LHC online and offline tracking

Steven **Farrell**[1,a], Dustin **Anderson**[2], Paolo **Calafiura**[1], Giuseppe **Cerati**[3], Lindsey **Gray**[3], Jim **Kowalkowski**[3], Mayur **Mudigonda**[1], **Prabhat**[1], Panagiotis **Spentzouris**[3], Maria **Spiropoulou**[2], Aristeidis **Tsaris**[3], Jean-Roch **Vlimant**[2], and Stephan **Zheng**[2]

[1] *Lawrence Berkeley National Laboratory, Berkeley, California, United States of America*
[2] *California Institute of Technology, Pasadena, California, United States of America*
[3] *Fermi National Accelerator Laboratory, Batavia, Illinois, United States of America*

Abstract. Particle track reconstruction in dense environments such as the detectors of the High Luminosity Large Hadron Collider (HL-LHC) is a challenging pattern recognition problem. Traditional tracking algorithms such as the combinatorial Kalman Filter have been used with great success in LHC experiments for years. However, these state-of-the-art techniques are inherently sequential and scale poorly with the expected increases in detector occupancy in the HL-LHC conditions. The HEP.TrkX project is a pilot project with the aim to identify and develop cross-experiment solutions based on machine learning algorithms for track reconstruction. Machine learning algorithms bring a lot of potential to this problem thanks to their capability to model complex non-linear data dependencies, to learn effective representations of high-dimensional data through training, and to parallelize easily on high-throughput architectures such as GPUs. This contribution will describe our initial explorations into this relatively unexplored idea space. We will discuss the use of recurrent (LSTM) and convolutional neural networks to find and fit tracks in toy detector data.

1 Introduction

The Large Hadron Collider (LHC) [1] at CERN is the most powerful particle accelerator and collider built on Earth to-date. It accelerates two beams of protons in opposing directions to near-light speed and collides them to produce particle products via proton-proton interactions. Experiments such as ATLAS [2] and CMS [3] use building-sized detectors to reconstruct these collision events and make statistical tests of the fundamental laws of nature. The detectors and their reconstruction software must identify the types and momenta of $O(10^3)$ particles per collision with high efficiency and precision.

One of the most essential aspects of collision event reconstruction at the LHC is charged particle tracking. Silicon-based, high-granularity tracking sensors detect ionization charge deposited by particles as they propagate through the detector in a magnetic field. Pattern recognition tracking algorithms and subsequent estimation methods use this information to measure the curvature of particle trajectories and thus deduce the particles' charge and momentum. This information is combined with other signals in the detector to reconstruct entire collision events.

[a] e-mail: SFarrell@lbl.gov

The High-Luminosity LHC (HL-LHC) [4] will bring considerable performance challenges to particle tracking when it will begin operations around 2026. The average number of interactions per beam crossing will increase by a factor of ten compared to current conditions and result in a similar increase in the number of charged particles and hence detector occupancy. Traditional tracking algorithms scale quadratically or worse with detector occupancy. Under reasonable assumptions on the evolution of CPU performance and cost, and assuming flat budgets for LHC computing resources, to process $O(50)$M charged particle tracks per second at the HL-LHC, tracking algorithms will need to become one order of magnitude faster and run in parallel on one order of magnitude more processing units (cores or threads).

The HEP.TrkX project is exploring the applicability of advanced machine learning algorithms to HL-LHC track reconstruction. We have started by studying deep neural network [5, 6] architectures, which in principle are well suited for the HL-LHC tracking problem, given their ability to learn effective representations of high-dimensional data through training, and to model complex dynamics through computationally regular transformations that scale linearly with the input data size and parallelize easily on high-throughput architectures such as FPGAs or GPUs.

2 Traditional tracking algorithms

The current generation of LHC charged particle track reconstruction (tracking) methods divides the problem into stages. These stages are hit clustering, track seed finding, track building, and track fitting. Development of these algorithms started in the 1990s and they have been continually optimized to reduce processing time. Nonetheless, the most CPU-intensive steps of this approach, seeding and track building, scale worse than quadratic with the number of position measurements and hence will become major CPU bottlenecks in the expected HL-LHC conditions. There are several excellent reviews of traditional LHC tracking strategies [7–9]. Here we will only give a brief overview of some relevant features of the algorithms.

Hit clustering involves estimating the charge deposit and position parameters of particles that intercept with multiple pixels on a detector layer. This transforms the discrete detector hardware space of $O(10^8)$ measurements into a continuous space of $O(10^5)$ space-points in 3D. Modern approaches use shallow neural-networks to identify and disentangle shared clusters [10].

In the track seed finding stage, sets of three space-points (one per detector layer) are combined to form triplets[1] [10, 11]. A triplet is the smallest set of space-points that allows to estimate the track curvature (hence the charge and momentum) and its perigee (point of closest approach) with respect to the center of the interaction region, assuming a helical trajectory. Good triplet seeds are selected by requirements on the triplet geometry, its momentum, and its perigee.

The resulting list of filtered triplets, or seeds, is used as input for the track building stage. A Combinatorial Kalman Filter [12] (CKF) is used to build track candidates from the chosen seeds. The filter creates multiple track candidates per seed if more than one compatible space-point extension exists on the same layer. The CKF is very efficient in finding tracks from charged particles originating from the collision region and in rejecting "fake" tracks created by random combinations of space-points.

To finish track reconstruction, a full resolution fit is performed on the final set of track candidates trajectories. At this stage, additional quality criteria may be applied to the track candidates along with some ambiguity resolution of shared space-points [10]. The result is a set of physics objects

[1] To be precise, a triplet can be composed of three space-points, or two space-points and a constraint coming from the beam-beam interaction region (beam-spot).

describing the trajectory parameters and momenta of the particles in the event. A typical physics event at the HL-LHC will have $O(10^3)$ reconstructed tracks.

3 Machine learning for tracking

Deep learning [5, 6] has not yet been explored in depth for the problems of particle tracking, so there are many open questions about the best way to incorporate such techniques.

What part of the problem to address? Possible candidates include hit clustering, seed identification and classification, and track building and fitting. Machine learning algorithms may also be able to combine steps, such as track building and fitting, or track finding without seeds by considering problems in an end-to-end manner which has proven successful in many other domains such as robotics [13].

How to represent the data? Though closely related to the previous question, this one is more data-oriented. A machine learning algorithm could use as input the clustered spacepoints which are continuously distributed in 3D space or could work with the discrete space of pixel sensors and learn its own intermediate representations. Similarly, the output of a track finding algorithm could follow the traditional approach of providing a strict grouping of spacepoints into tracks with minimal sharing between track candidates, or it could give a soft-assignment of spacepoints to tracks, and/or it could give track trajectory parameters directly. These choices may have significant tradeoffs in terms of algorithm simplicity, performance, interpretability, and overall usefulness to experiments.

How to address the many other challenges? These include sparsity and irregularity of the tracking data, difficulty in defining differentiable cost functions, and experiment requirements for fine-level control and interpretability of the model. Finally, any viable models should respect the time and memory budget constraints of the experiment.

In this paper we present some studies on just a couple of possible approaches that we believe to be promising. The first, described in section 5, is seeded single-track building with recurrent and convolutional models. The next, described in section 6, is end-to-end track parameter estimation using convolutional + recurrent networks.

Recurrent neural networks like Long Short Term Memory (LSTM) [14] networks can be used as state estimators in a way that is similar to a Kalman Filter. Detector layers can be fed into the model as an input sequence of pixel arrays. The model uses these measurements to update its hidden state and provides a prediction for the target detector layer. The target layer can be the same layer as the input or it can be the next detector layer in the case of a forward-predictor type of model. The output of the network can be used to assign hits to a single track or possibly to multiple tracks simultaneously.

Convolutional networks are expected to be useful for pattern-recognition in track finding. Tracking data exhibits some of the traits of natural images that are naturally exploited by convolutional networks such as translational symmetries, locality, and hierarchical multi-scale structure. In CNNs for natural images, the first convolutional layers are commonly trained to detect low-level features like edges, so it is expected that the initial layer of a CNN for tracking would identify stubs of compatible hits in adjacent layers. Later layers of the network would then connect stub features together to form track segments, and so on until a model of an entire track or set of tracks is constructed. This conceptual model, which is useful only for intuition and may not reflect the actual learned behavior of a convolutional network, is illustrated in figure 1.

More advanced methods may also be applicable to tracking problems. For example, one can view the track finding problem as similar to the image-captioning problem [15], where event data of hits represents the images and descriptions of the tracks (hit assignments or parameters) are analogous to the text captions. We make use of such a model in section 6. Such models can be augmented with

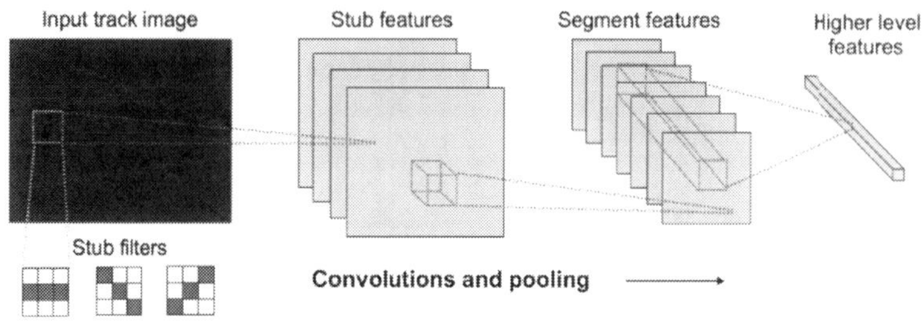

Figure 1. A possible interpretation of convolutional neural networks applied to 2D tracking data.

attention mechanisms [16–18], giving the models the capability to focus on particular parts of the input or intermediate feature representations to produce a desired output.

Such rich learned representations have also proven highly beneficial for tracking-based problems in non-HEP applications, such as sports analytics and computational neuro-science. For instance, [19] uses attention-based LSTMs to learn hierarchical models of basketball player behavior from tracking data, while [20] applies recurrent neural networks to generate realistic fruit-fly behavior and handwriting.

4 Datasets

Simple toy datasets were used to study and demonstrate the ideas discussed in this paper. The "detectors" are made of perfect pixel planes in 2D or 3D. Tracks are sampled from straight lines contained within the detector volume, and binary hits are recorded in each intercepting discrete pixel on each layer. No trajectory curvature, material effects, or detector inefficiencies are modeled. These toy datasets are highly simplistic compared to real tracking detector data, which means that quantitative results are likely not indicative of algorithm performance in realistic scenarios. Nonetheless, this simple toy data provides a useful environment to test out various models. Figure 2 shows example 2D data generated with tracks as well as uniform noise. Figure 3 shows an example 3D event.

For the experiments described in section 5, the following data configurations were used. 2D toy experiments used one million 2D events with 50 detector layers of 50 pixels each, one signal track, and five background tracks for training. The 3D toy experiments used a detector with 10 layers and 50×50 pixels in each layer. Events were generated with a random number of background tracks sampled from a Poisson distribution with mean values varied from 1 to 100. At each point, five million events were generated for training and one hundred thousand events for testing.

5 Track finding with LSTMs and CNNs

The goal of this line of study is to identify models which can do the assignment of pixel hits to a track candidate by extrapolating from a partial track (a seed) through detector layers. When considering a single track at a time, the problem can be formulated as one of multi-class classification. The pixels in one detector layer make up the possible "classes", and the model must identify which one is traversed by the target track candidate. Modeling of track dynamics can be handled by LSTMs or CNNs.

A basic LSTM model for 2D track finding is shown in figure 4. This model consists of an LSTM layer which reads the input pixel arrays and a single fully-connected layer which is applied separately to each LSTM output to produce the pixel predictions for the same detector layer. The seed is specified

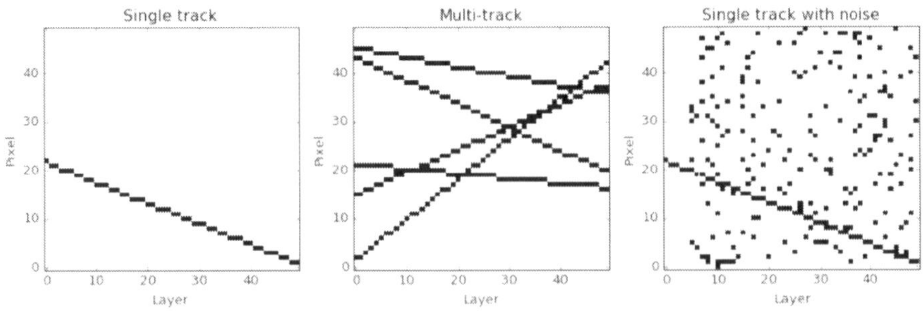

Figure 2. 2D plane detector toy data with a configurable size, straight line tracks, and optional uniform noise hits.

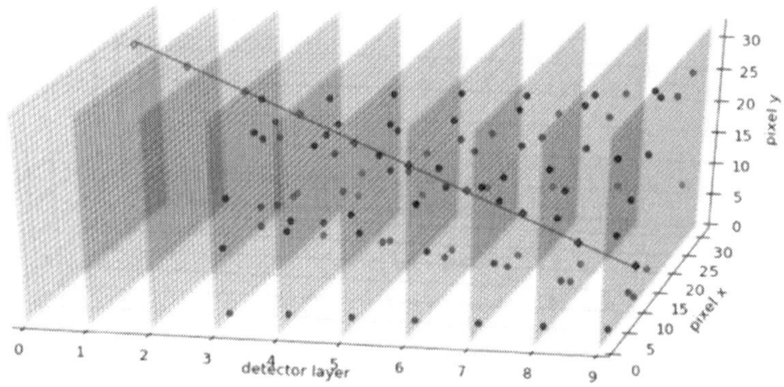

Figure 3. 3D plane detector toy data with ten detector layers of configurable size, straight line tracks, and optional uniform noise hits.

in the input by zeroing-out all non-candidate hits in the initial seed layers. Figure 5 shows an example model input and prediction after training on events with one target track and five background tracks. The simple LSTM architecture is easily able to reconstruct the target track, indicating that it can disentangle the combinatorics at each detector layer.

In most real tracking detectors, including those at the LHC, the number of pixels in each detector layer is not fixed and in fact increases with each successive layer. To adapt the simple rectangular image model to such a variable-sized data structure there are some options. One straightforward approach is to pad the detector image with zeros to recover the rectangular image. Another approach is to transform each pixel array through a dedicated fully connected layer to a fixed size hidden representation which is processed by the LSTM as before. In such a model, dedicated fully-connected layers on the output side of the network then transform the fixed size representation back to the target detector layer size. Figure 6 shows an example where toy data was simulated in the full rectangular detector shape but a wedge shape was cut out to demonstrate the variable-length layer model.

Convolutional neural networks can be applied to the exact same problem. In this case, the inputs and outputs are formatted in the same way, but instead of iterating through the detector layers as a sequence, we apply several layers of convolutional filters to the detector image. For the final prediction output a softmax is again applied over the pixels in each individual detector layer. Figure 7 shows what

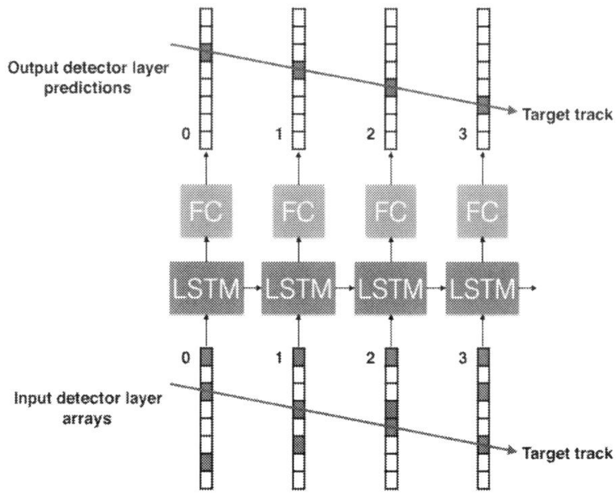

Figure 4. The basic LSTM model architecture used to classify hits for one track. The LSTM and a fully-connected layer with a softmax activation read the pixel arrays and predict which pixels belong to the target track.

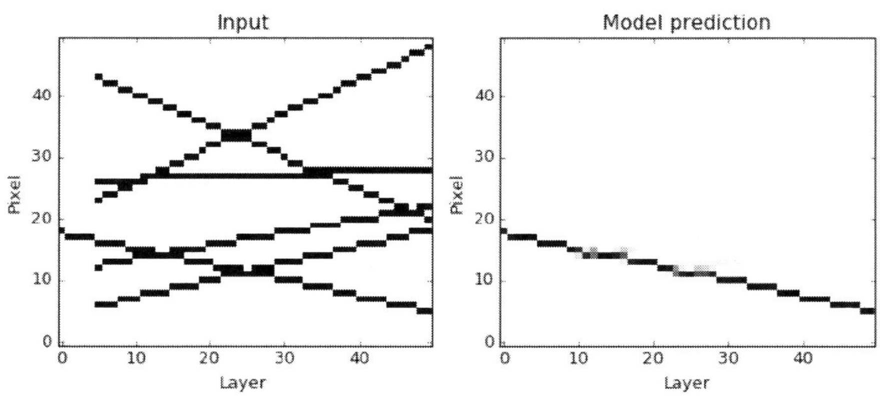

Figure 5. LSTM model input (left) and prediction (right) for an example with one target track and five background tracks. The first five detector layers are used to specify the target track seed. In the prediction, darker values mean greater confidence.

happens when a ten-layer convolutional model with no down-sampling and filter size (3×3) is applied to a fifty-detector-layer toy dataset. The extrapolation reach of the model is limited by the number of convolutional layers, illustrating a potential limitation of this kind of architecture. One way to combat this is by introducing pooling layers which down-sample the data and allow to combine information across the entire image. It is then necessary to introduce up-sampling and additional convolutional layers to recover the track prediction at the original image size. This is essentially an autoencoder model which tries to reconstruct a subset of its input. An example input and output after training such a model is shown in figure 8.

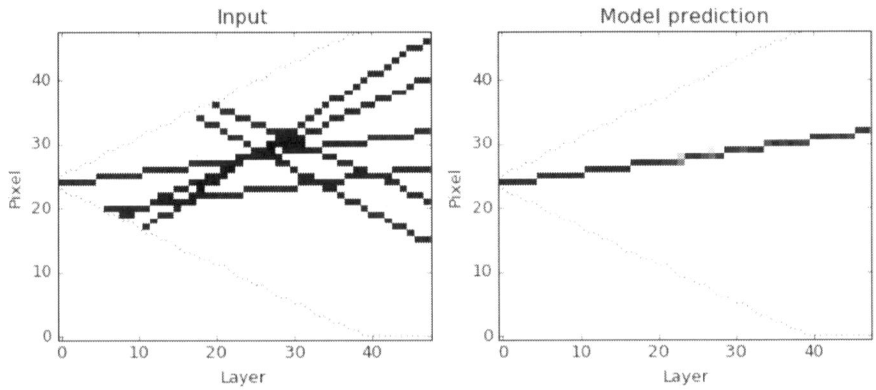

Figure 6. LSTM model input (left) and prediction (right) for an example with variable-sized detector layers. The white edges define the wedge detector shape as fed into the model.

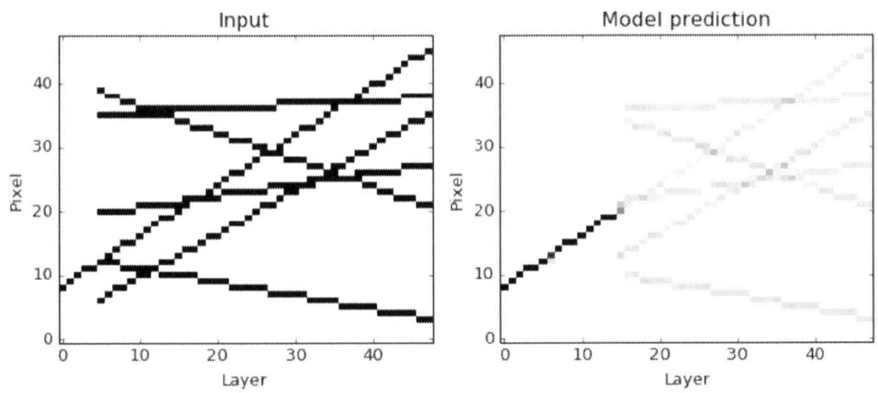

Figure 7. Input and model prediction of a simple 2D convolutional model with ten convolutional layers and no down-sampling. The extrapolation of the seed track is limited in this case because the architecture does not allow to communicate information across all detector layers.

Extending the models to 3D toy detector data is mostly straightforward. For the LSTM model, all that's required is to flatten the 2D detector layer pixel arrays into 1D arrays and proceed as before. The convolutional models need to adopt 3D convolutions. For a more detailed study of this more interesting type of toy data, we consider additional variations on the previously mentioned models. A deep-LSTM model is the same as the original LSTM model but adds additional fully-connected layers before and after the LSTM. A bidirectional LSTM model (or biLSTM) uses an additional LSTM layer which processes the detector layers in reverse order. The two LSTM outputs are concatenated before being fed into the fully connected layers for the final prediction. A next-layer LSTM model makes forward predictions on the next detector layer before reading the data on that layer.

The accuracies of these models were measured on the 3D experiment datasets with the mean number of background tracks varied from 1 to 100. Figure 9 shows the test-set classification accuracies

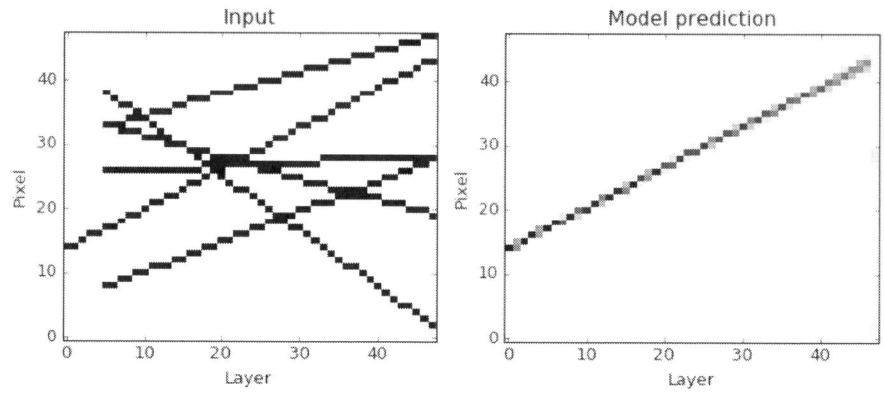

Figure 8. A convolutional autoencoder model. Down-sampling between convolutions allows the model to connect information across the entire image, and up-sampling allows to recover the track of interest in the original image size.

calculated in two ways: considering all pixels on each layer and considering only the pixels with hits on each layer. An example event and model predictions for the basic LSTM and the next-layer LSTM are shown in figure 10. Interestingly, the next-layer LSTM is one of the worst performers at picking the correct pixel, but is the most accurate at classifying the pixels that actually have hits. This is presumed to be because the forward-thinking nature of the model gives broad, smooth track predictions that work well on straight tracks. Another interesting feature is that the convolutional models seem to scale better to high track multiplicity. Finally, none of the models perform particularly well at high track multiplicity, showing unsatisfying hit classification accuracies around 60% for 100 mean number of background tracks.

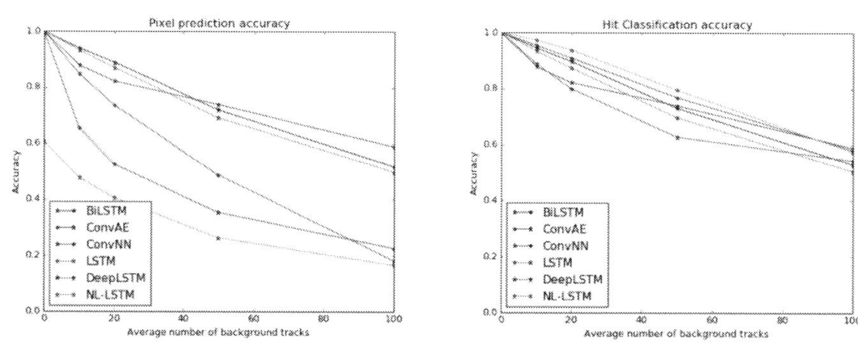

Figure 9. Pixel (left) and hit (right) classification accuracies for various models on the 3D toy data. The pixel accuracy is calculated per detector layer by comparing the top scoring pixel to the target pixel. The hit accuracy is calculated per detector layer by choosing the top scoring pixel that has a hit and comparing to the target pixel.

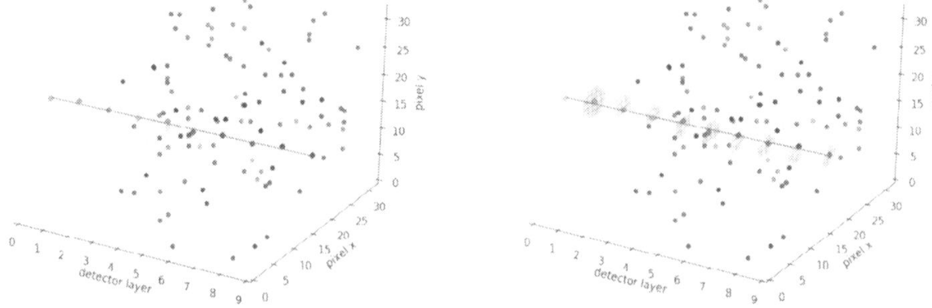

Figure 10. Example displays of a 3D toy event and model prediction for the basic LSTM (left) and the next-layer LSTM (right). The target track is shown in red. Prediction pixels are only shown for scores larger than 0.01. The basic LSTM prediction precisely follows the target hits, whereas the next-layer LSTM gives broad predictions due to it's forward-predicting nature.

6 Track parameter estimation with LSTMs and CNNS

In addition to the hit classification models discussed in the previous section, we also considered end-to-end models that process the raw detector data without track seeds, identify tracks in the input data, and estimate the parameters of each track. The outputs of these models could be useful on their own as quantities for physics usage, or they could be used as supplementary training targets in the classification-of-hits approach.

Figure 11 shows a model, built from CNNs and an LSTM, that can process a detector image and emit parameters for a variable number of tracks. In the case of the toy detector data considered here, the track parameters correspond to a line's slope and intercept. A CNN with four layers, with a max-pooling operation between the second and third layers, is sufficient to extract relevant features from the input image. An LSTM stacked on top of the convolutional layers acts as a simple attention mechanism; it emits track parameters for each track in sequence, moving from smallest intercept value to largest. The performance of the model for events with between one and six tracks is illustrated in Figure 12, which shows the distribution of the difference between the true and predicted slope and intercept values.

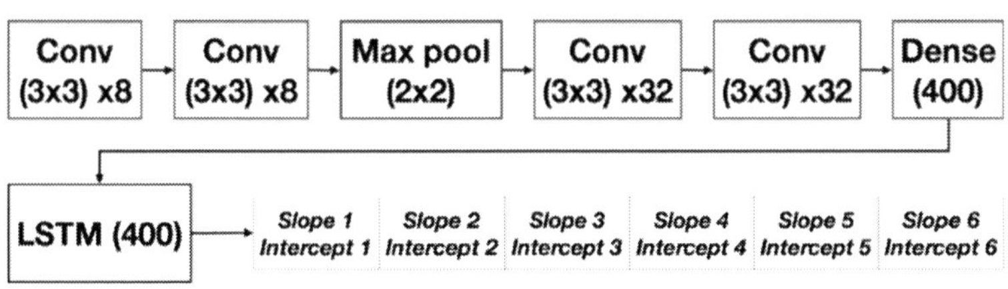

Figure 11. Diagram of a model with convolutional layers followed by dense-fully-connected and LSTM layers to predict track parameters for a variable number of tracks.

Uncertainties on the predicted track parameters can be obtained via an additional model output that represents the track covariance matrix. To obtain uncertainties that reflect the model's true per-

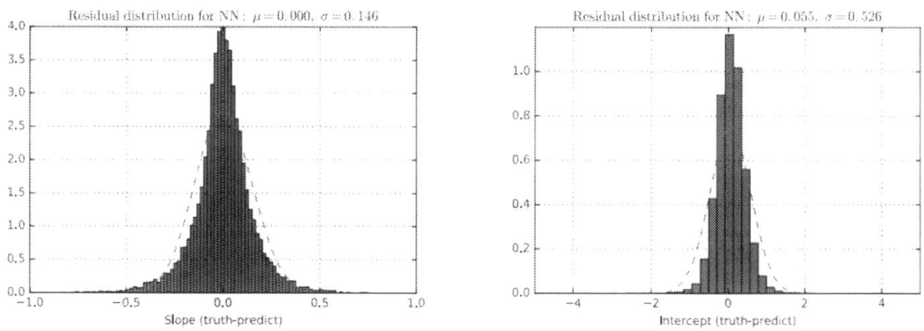

Figure 12. Difference between true and predicted values of track slope (left) and intercept (right) for the CNN+LSTM model shown in Fig. 11. The distributions are constructed using events with up to six tracks.

formance, the model is trained by maximizing the following log gaussian likelihood, which relates the true track parameters y, their predicted values $f(x)$, and the covariance matrix Σ output by the network:

$$L(x, y) = \log|\Sigma| + (y - f(x))^T \Sigma^{-1} (y - f(x)) \tag{1}$$

Figure 13 illustrates the predicted track parameters and their uncertainties for an example event with six tracks.

The convolutional filters learned by the model provide insight into the features that the model has learned to extract. A filter can be visualized by using gradient ascent to find an input image that maximizes the filter's output activation [21]. Using this technique we generated the images shown in figure 14, which provide visualization of the filters in the second convolutional layer of the CNN+LSTM model.

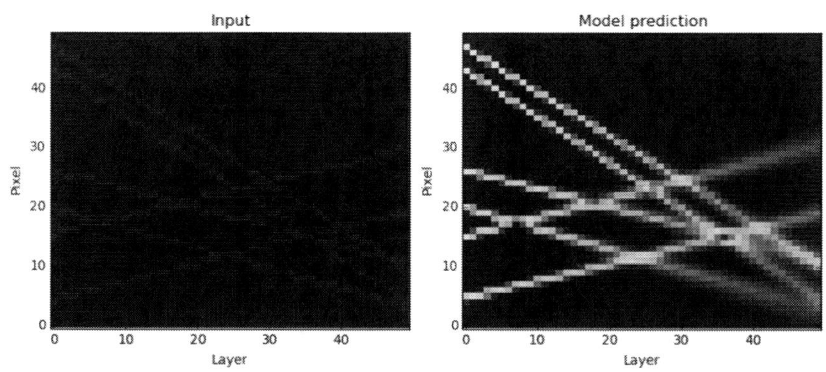

Figure 13. Example 2D toy detector event (left) with a graphical illustration of the output predictions (right) for the straight line track parameters and covariances. The prediction is generated by sampling track parameters many times according to their covariance matrix and plotting the resulting lines.

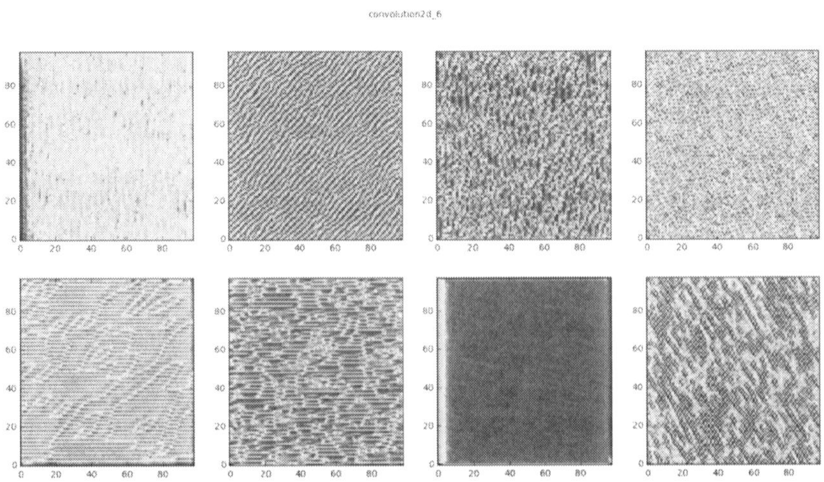

Figure 14. Visualization of the CNN+LSTM model's second layer filters via gradient ascent optimization of the inputs.

7 Conclusion

In this paper we have prototyped and demonstrated some preliminary ideas for adopting machine learning techniques for particle tracking problems. This is still a relatively unexplored space and we have only just begun to scratch the surface, but nonetheless some of the technologies used in the field of deep learning show some promise in this field as well. LSTMs were seen to be effective at the problem of hit assignment in 2D and 3D with a sequence of detector layer measurements and may be viable as an alternative to the combinatorial Kalman Filter. CNNs were shown to be able to construct representations of the detector data in a ground-up fashion that are useful for hit assignment or parameter and uncertainty estimation. By attaching an LSTM to a CNN model we demonstrated a potentially powerful end-to-end model which can find a variable number of tracks in a detector image. All of the explored models were trained and tested on GPUs, demonstrating parallelization capabilities that help to address the scaling issues of traditional tracking algorithms.

There are numerous clear hurdles to adopt these methods in realistic scenarios. Real particle detectors are typically irregular in layout and may not be easily transcribed into the image format. The high granularity and sparsity of real detector data poses a dimensionality and scaling problem for solutions that work in discrete spaces.

Our proposals for areas of future study build off the work performed. First and foremost, the ideas discussed need to be scaled up to greater realism in terms of detector size, granularity, and track multiplicity. The ACTS [22] software package will be a useful tool for fast simulation and digitization of data that is semi-realistic with a generic LHC-like detector. Second, a major missing piece from the work shown so far is comparison with traditional algorithm solutions. The combinatorial Kalman Filter is the proper algorithm for the pattern recognition problem, and similarly, KF-based track fitting methods for track parameter estimation. Finally, we believe there are numerous areas of research in developing new ideas to solve these problems. Machine learning models that can work in the continuous space of clustered hits may be more appropriate for handling the data sparsity and irregularity of detector geometry.

All models studied in this paper were implemented in the Keras [23] framework. The code with full details of the models and training in Jupyter notebooks is available at [24].

The authors would like to thank the funding agencies DOE ASCR and COMP HEP for supporting this work, as well as the numerous tracking experts from ATLAS and CMS who have shared insights and experience.

References

[1] L. Evans, P. Bryant, JINST **3**, S08001 (2008)
[2] G. Aad et al. (ATLAS), JINST **3**, S08003 (2008)
[3] S. Chatrchyan et al. (CMS), JINST **3**, S08004 (2008)
[4] B. Schmidt, J. Phys. Conf. Ser. **706**, 022002 (2016)
[5] Y. LeCun, Y. Bengio, G. Hinton, Nature **521**, 436 (2015)
[6] I. Goodfellow, Y. Bengio, A. Courville, *Deep learning* (MIT Press, 2016)
[7] R. Fruehwirth, M. Regler, R. Bock, H. Grote, D. Notz, *Data Analysis Techniques for High-Energy Physics*, 2nd edn. (Cambridge University Press, Cambridge UK, 2000)
[8] F. Ragusa, L. Rolandi, New Journal of Physics **9**, 336 (2007)
[9] R. Frühwirth, A. Strandlie, *Pattern recognition and reconstruction: Datasheet from landolt-börnstein - group i elementary particles, nuclei and atoms · volume 21b1: "detectors for particles and radiation. part 1: Principles and methods" in springermaterials (http://dx.doi.org/10.1007/978-3-642-03606-4_13)*, copyright 2011 Springer-Verlag Berlin Heidelberg, `http://materials.springer.com/lb/docs/sm_lbs_978-3-642-03606-4_13`
[10] The ATLAS Collaboration, *The Optimization of ATLAS Track Reconstruction in Dense Environments* (2015), `https://cds.cern.ch/record/2002609`
[11] A. Salzburger, *Tracking at HL-LHC and FCC* (2015), `https://indico.hephy.oeaw.ac.at/event/86/session/2/contribution/4`
[12] R. Mankel, Nuclear Instruments and Methods in Physics Research Section A: Accelerators, Spectrometers, Detectors and Associated Equipment **395**, 169 (1997)
[13] S. Levine, C. Finn, T. Darrell, P. Abbeel, arXiv preprint arXiv:1504.00702 (2015)
[14] S. Hochreiter, J. Schmidhuber, Neural Comput. **9**, 1735 (1997)
[15] O. Vinyals, A. Toshev, S. Bengio, D. Erhan, CoRR **abs/1411.4555** (2014)
[16] M. Denil, L. Bazzani, H. Larochelle, N. de Freitas, CoRR **abs/1109.3737** (2011)
[17] V. Mnih, N. Heess, A. Graves, K. Kavukcuoglu, CoRR **abs/1406.6247** (2014)
[18] K. Xu, J. Ba, R. Kiros, K. Cho, A.C. Courville, R. Salakhutdinov, R.S. Zemel, Y. Bengio, CoRR **abs/1502.03044** (2015)
[19] S. Zheng, Y. Yue, J. Hobbs, *Generating Long-term Trajectories Using Deep Hierarchical Networks*, in *Advances in Neural Information Processing Systems* (2016), pp. 1543–1551
[20] E. Eyjolfsdottir, K. Branson, Y. Yue, P. Perona, arXiv preprint arXiv:1611.00094 (2016)
[21] K. Simonyan, A. Vedaldi, A. Zisserman, CoRR **abs/1312.6034** (2013)
[22] *A Common Tracking Software Project*, `http://acts.web.cern.ch/ACTS/index.php`
[23] F. Chollet et al., *Keras*, https://github.com/fchollet/keras (2015)
[24] *HEP.TrkX CTD2017 public code repository*, `https://github.com/HEPTrkX/heptrkx-ctd`

L1 track trigger for the CMS HL-LHC upgrade using AM chips and FPGAs

Giacomo **Fedi**[1,a]

[1] *Università di Pisa and INFN Pisa, Italy*

Abstract.
The increase of luminosity at the HL-LHC will require the introduction of tracker information in CMS's Level-1 trigger system to maintain an acceptable trigger rate when selecting interesting events, despite the order of magnitude increase in minimum bias interactions. To meet the latency requirements, dedicated hardware has to be used. This paper presents the results of tests of a prototype system (pattern recognition mezzanine) as core of pattern recognition and track fitting for the CMS experiment, combining the power of both associative memory custom ASICs and modern Field Programmable Gate Array (FPGA) devices. The mezzanine uses the latest available associative memory devices (AM06) and the most modern Xilinx Ultrascale FPGAs. The results of the test for a complete tower comprising about 0.5 million patterns is presented, using as simulated input events traversing the upgraded CMS detector. The paper shows the performance of the pattern matching, track finding and track fitting, along with the latency and processing time needed. The p_T resolution over p_T of the muons measured using the reconstruction algorithm is at the order of 1% in the range 3-100 GeV/c.

1 Introduction

For the High Luminosity (HL) upgrade of the Large Hadron Collider (LHC), scheduled for 2026, the number of proton-proton interactions per bunch crossing (pile-up) is predicted to reach 140 at the design instantaneous luminosity of 5×10^{34} cm^{-2}s^{-1}. With an increase of $\approx$5 times of the pile-up count with respect to the current LHC operations, the CMS trigger system needs to be redesigned to improve its performance.

CMS's trigger system is divided in two stages. The first stage, called level-1 or L1 trigger, decides whether or not to save the information coming from the detector within few microseconds after the bunch crossing (event). This system is mainly implemented using hardware and firmware logic. The second step is called the High Level Trigger (HLT), and makes a further decision on the level-1 accepted events. While the HLT system can use all the available event information, the L1 trigger makes its decision based on partial and imprecise information.

Currently, the L1 trigger does not use the tracker information in its decisions. A way to mitigate the increase of pileup could come from the use of high-resolution spatial information from the silicon tracking detectors, at the cost of adding a few microseconds of latency and with a large increase in

[a] e-mail: giacomo.fedi@cern.ch

the amount of data that needs to be processed compared to the current L1-trigger. The trajectories of charged particles in CMS are helices since they are bent by the magnetic field generated by a solenoid. When a charged particle passes through the silicon tracker it deposits a small amount of energy in the silicon sensors and the passage of the particle is recorded ("hit"). The CMS tracker for the HL-LHC upgrade will be composed of six (five) layers of silicon modules in the central (forward) region. Each proton-proton interaction and their interaction with the detectors can generate hundreds of charged particles so that the silicon tracker generates thousands of hits in each event. The large number of hits makes the reconstruction of the trajectory of charged particles in few microseconds very challenging.

For the HL phase of the LHC, the outer tracker of CMS will be redesigned, adopting a novel sensor technology to reduce the event information that is needed to be transferred from the detector to the data acquisition system. Each outer tracker layer will be composed of modules called "p_T modules", and each module is made of pairs of silicon sensors. The p_T module concept relies on the fact that the strips of both sensors are parallel to the beam axis in the barrel and nearly radial in the endcap, and uses the correlation of signals in closely-spaced sensors (a few millimeters) to make an angular measurement of the pair of hits. These pairs of hits are called "stubs". Using the stub angular information and assuming the tracks are coming from the interaction volume, the p_T of the track that is generating the stub can be measured. By imposing a $p_T > 2$ GeV/c requirement, a reduction by O(10) in the amount of data that needs to be transmitted to the trigger can be achieved.

CMS is exploring three different approaches to tackle the track reconstruction at L1 trigger. Two are based on pattern matching and track reconstruction done with FPGAs [1, 2]. The third, to be discussed in this report, is based on the use of an associative memory (AM) ASIC for the pattern matching and FPGAs [3] for the track reconstruction. The purpose of our R&D activity is to prove that CMS's tracking requirements can be met by combining pattern recognition based on AM custom chips and computing capabilities available in state-of-the-art FPGA devices. Track finding based on Associative Memories [4] has been successfully used in the CDF experiment [5] and, more recently, is being exploited in the Fast-Track processor [6, 7] for the ATLAS Level-2 trigger system.

In the AM plus FPGA approach, the CMS outer tracker will be divided in 6×8 regions in $\eta \times \phi$ (pseudo-rapidity and azimuthal angle) called "trigger towers". In each trigger tower, the track finding will be performed using data from p_T modules belonging to that region. Each tower will receive an average of $\approx$100 hits per layer for each bunch crossing. We built a prototype of the system with currently available technology to demonstrate that we will be able to build the full system for the HL phase of the LHC. Our demonstration requires only a small extrapolation of the ASIC technology and cost of commercial devices, and depends on the assumption of a strong R&D program on the AM chip.

2 Demonstrator System Overview

A small prototype (demonstrator) of the system has been built to prove that the AM plus FPGA approach can satisfy CMS' HL-LHC requirements. The demonstrator is composed of two ATCA shelves; one ATCA shelf (data-source ATCA) is emulating the data output of the CMS outer tracker, while the second ATCA shelf receives the data and is equipped with the ATCA boards and mezzanine boards that reconstruct the tracks. The two ATCA shelves are physically connected by 480 10 Gbps fiber-optic links, leading to a maximum data rate of 4.8 Tbps.

Both ATCA shelves are equipped with several instances of ATCA custom boards, called Pulsar IIb [8]. These boards were designed at FNAL for data delivery in high energy physics experiments. Two types of PRM have been developed: one with a full AM chip configuration and one with two Xilinx Ultrascale FPGAs and a socket which can accommodate VIPRAM [9] chips. Each Pulsar IIb

Figure 1. Picture showing a part of the system. An ATCA shelf accommodates ten Pulsar IIb boards. Each board can accommodate two PRMs.

board can accommodate two Pattern Recognition Mezzanine (PRM) boards; these boards perform the pattern recognition and track fitting. Figure 1 shows a schematic view of the track reconstruction system based on ATCA technology.

Each PRM board contains one or two Xilinx FPGAs, implementing data flow management and track fitting, and one or multiple AM devices. Each PRM board receives full resolution data and temporarily stores them in a data-buffer. Full resolution data are the coordinates of the lower sensor cluster and the angle information. Lower resolution data, called SuperStrip ID (SSID), are evaluated and transmitted to custom chips where the AMs are implemented (AM chip). The AM chip acts as a Content Addressable Memory (CAM) and quickly matches the SSIDs with the pre-loaded patterns. Pre-loaded patterns are based on simulation of the charged particles originating from the proton-proton collision and traversing the tracker. The tracks mostly originate in a well-defined region of the detector and the curvature of the helices is restricted to a certain region of interest. p_T modules are able to discriminate stubs associated with track p_T above 2 GeV/c, but in the current implementation of the prototype, p_T>3 GeV/c threshold has been explored instead, in order to simplify the prototype.

All the patterns present in the AM chip are simultaneously compared with the input data for each layer. When a match is found on a layer, a match bit is set and it remains set until the end of the event processing, when an internal reset signal is propagated. If in a given pattern a match is detected for all the layers, or for a majority of them, the pattern address is transmitted to the FPGA. In the current implementation only one missing layer is allowed. The full resolution data associated with these SSIDs are retrieved from the FPGA data-buffer, filtered and propagated to the Track Fitting module which fits the set of stubs with the same majority set in the AM chips. The processing time for each event in a PRM board is of the order of few μs, while each PRM board receives events with a period of 500 ns. Therefore, it is necessary to time-multiplex the data to 10 Pulsar IIb boards using a round-robin mechanism, exploiting the full mesh backplane of the ATCA shelf. Tracks that are found in the PRM will then be transferred to the Global Trigger System [10]. The final goal of the prototype board is to evaluate the performance of the real-time system described above using state-of-the-art AM chips. The prototype is also used to extrapolates the future needs, in terms of system requirements, to match the processing time constraint of 4-5 μs.

For this purpose, two prototype processors have been developed: the Pattern Recognition Mezzanine PRM06 at INFN and the FNAL PRM. The processors are designed to cope with the pattern recognition task [11] using the new version of the AM device. The track parameters are then measured for each pattern using a precision track fit [7] performed with the latest generation of FPGA devices.

 DOI: 10.1051/epjconf/201715000004

Figure 2. Left: INFN pattern recognition mezzanine board. Right: FNAL pattern recognition mezzanine board.

3 PRM architecture

The PRM board is shown in Figure 2. It is a custom 14.9×14.9 cm^2 printed circuit board housing the following components:

- FMC connectors (2x);
- Xilinx Kintex Ultrascale FPGA (1x INFN PRM, 2x FNAL PRM);
- Associative Memory devices (12x INFN PRM, 1x FNAL PRM);
  - FNAL AM: FPGA emulated chip with 4k patterns;
  - AM06: 65 nm custom chips with 128k patterns each;
- Power regulators; and
- DDR Pattern Memory (2x INFN PRM, 1x FNAL PRM).

The two FMC connectors are used to interface the PRM to the Pulsar IIb host board. They carry signals and power supply voltages. The host board provides 3.3 V and 12 V, for a maximum available power of 150 W for each PRM. The power regulator group is used to generate the 1.0 V, 1.2 V, 1.8 V and 2.5 V voltages required by the FPGA, the AM devices and the pattern memory. Six bidirectional high-speed serial links (up to 10 Gbps, limited by the FMC connectors), with three links through each FMC connector, are used to send and retrieve data from the PRM with a total input and output bandwidth of 60 Gbps. Moreover, 68 additional LVDS pair pins are used for monitoring and to provide additional bandwidth for data transfer.

The Xilinx Kintex Ultrascale FPGA manages the data flow inside the mezzanine. It receives and temporarily stores the full resolution hit data from the Pulsar IIb board and evaluates and distributes the lower-resolution hit data used in the pattern recognition to the AM chips. Finally, the external pattern memory provides up to 576 Mbit for INFN PRM and up to 36 Mbit for FNAL PRM of memory resources to store a copy of the pattern banks of the AM chips.

The FNAL PRM is equipped with two Xilinx Ultrascale 040/060 FPGAs. Fast serial links connect the two FPGA to each other with a maximum data rate of 16.3 Gbps. One of the FPGA is also fully connected with six high-speed serial links to the FMC connectors. The PRM also has additional external communication links consisting of QSFP+ transceivers.

Figure 3. Picture showing an ATCA shelf containting a Pulsar IIb board (leftmost slot) equipped with a PRM06, and an ATCA switch board (center slot).

The INFN PRM board houses 12 AM custom chips version 06 (AM06). Each chip can store up to 128k patterns, corresponding to a total capacity of 1.5M patterns per PRM. Each pattern is made of 8 independent 16-bit words, one for each layer. The input data bus distribution from the FPGA to the AM chips uses stages of fan-out buffers to distribute data in parallel to all the AM chips. SSIDs can be fed to the AM chips using two independent input buses. Each input bus is connected to a set of six AM chips. The input serial links can sustain a data rate of 2 Gbps, while the output data of each AM chip is connected directly to the FPGA using high speed serial lines running at 2.4 Gbps. The direct connection between the FPGA and the AM chips has been chosen to eliminate daisy chains and to reduce as much as possible the latency of the Level-1 trigger decision and to integrate all the functionalities in one single board.

The FNAL PRM board has a socket which can accommodate the soon-to-arrive VIPRAM [9] chip. One of the FPGAs is used to emulate the AM chip using synthesized HDL code, while the other FPGA is used to store the hits and to fit the matched patterns. The emulated AM chip can store up to 4k patterns. This mezzanine cannot manage the full number of patterns required for a trigger tower, but is designed to minimize the latency of the pattern-finding process. To this end, the FNAL PRM is equipped with a powerful FPGA with high speed serial links with data rates up to 16.3 Gbps, and the synthesized FPGA code has been designed to minimize latency. The aim of this PRM is to show what we can achieve with the future AM custom chips.

4 PRM hardware validation

Both INFN and FNAL PRM boards were validated regarding the electrical circuitry and both showed the expected behavior. The INFN PRM board was tested with a test stand consisting of a PRM06 board, an adapter card, and an evaluation board. An HTG Virtex 6 evaluation board was used for the main tests reported in this text, but electrical and logic connections were also tested using a Pulsar IIb board.

The HTG evaluation board has limited high-speed lanes to the FMC connectors. Only four high speed serial links are connected and the data rate is limited to 6.2 Gbps. To test the fast serial links of the INFN PRM, an Ultrascale Evaluation board was used. The evaluation board is connected to a host PC via Ethernet. It provides power, LVDS parallel connections and high-speed serial connections to

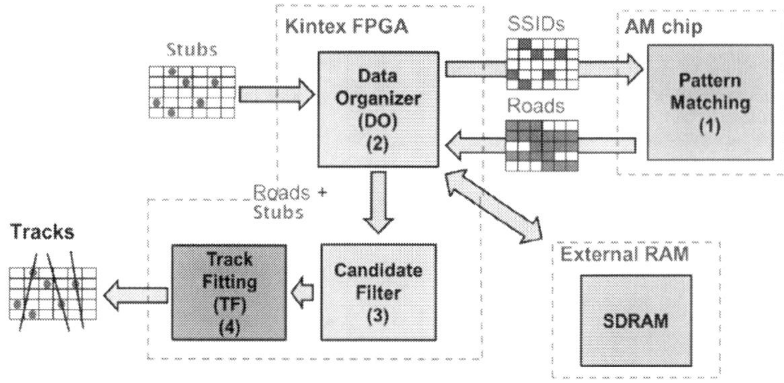

Figure 4. Schematic view of the modules of the track finding algorithm.

the PRM board. The IP-bus protocol is used to access devices in the Virtex 6 and in the PRM FPGAs (e.g.: control and status registers, source FIFOs containing test hits, monitoring FIFOs containing PRM outputs).

All the serial links between the FPGAs and to/from the AM chips on the PRM board were successfully tested. JTAG connections to the AM chips (2 JTAG chains with 4 AM chips each, and 2 JTAG chains with 2 AM chips each) have been tested to verify the ability to configure and program the AM chips. The eight input links to each AM chip were characterized. Pseudo Random Bit Sequences (PRBS) were used to test the links between the FPGA and the AM chips. The FMC connector links have undergone tests by using a loopback card and the IBERT tool, provided by Xilinx, that allows measuring the Bit Error Ratio (BER) and producing the corresponding eye diagrams.

Signal integrity up to 12.5 Gbps was tested by measuring the BER and the corresponding eye-scans in the receiver links of the PRM and in the receiver links of the Ultrascale board. The BER has been measured using a PRBS-7 sequence. The measured BER was less than 1×10^{-15} in both directions. The power consumption of the board has been measured by powering and configuring the four groups of the AM chips, resulting in a static power consumption of about 40 W. The mezzanine board has been successfully integrated with the Pulsar IIb board; see Figure 3. During these tests, the mechanical and electrical compatibility have been verified and both the LVDS and high-speed differential pairs were fully tested up to 10 Gbps. The two external DDR memories (RLD3RAM) were successfully tested using a dedicated Xilinx IP core.

The FNAL PRM board has been tested in a standalone configuration and with the Pulsar IIb board. The high speed serial links between the two FPGAs and towards the FMC connectors have been tested; the links between the FPGAs were validated at a rate of 16.3 Gbps, while towards the FMC connectors they were validated at a rate of 10 Gbps. The LVDS links between the FPGAs and the FMC connectors were successfully validated. The QSFP+ transceivers were also tested and no errors were found in the transmitted pseudo-data.

5 PRM functionalities

On both INFN and FNAL PRM boards, a synthesized HDL code has been developed to implement functionalities needed for the track finding: hit buffering, hit retrieving, and hit fitting. The tracking algorithm implemented in both PRM boards consists of the following sequence of operations (see Figure 4):

1. Full resolution stub data received by the host board are decoded and stored in a smart data buffer, called Data Organizer (DO), and coarse resolution stub data, SSIDs, are generated and transmitted to the AM chips;

2. AM chips perform pattern recognition on SSID input data and identify sets of SSID matching patterns, called "Road", that are transmitted back to the FPGA;

3. Roads are used to retrieve the associated SSID from an external pattern memory;

4. Stub data stored in the Data Organizer and belonging to the SSID retrieved from the pattern memory are propagated to the filtering module or to the combination builders;

5. The Track Fitter performs the fitting algorithm using the full resolution data from the Data Organizer to evaluate the track helix parameters and the goodness of the fit (χ^2).

Two different track finding algorithms were developed. Each one was designed to fulfill a different goal . In one case the algorithm focused on the reduction of the track finding latency; it is hosted by the FNAL PRM board. In the other case the algorithm focused on the handling of large pattern banks and multiple AM devices; it is hosted by the PRM06. The full chains of track finding have been developed and tested using HDL simulations and the actual FPGAs for simulated events.

The PRM06 reconstruction algorithm was intended to demonstrate the validity of the PRM design with multiple AM chips working in parallel. The PRM06 boards were equipped with the AM06 chip, which was not developed for stringent latency requirements. The latency of the pattern matching due to the AM06, about 1 μs, exceeds the 500 ns time period which needs to be met to manage the incoming data with the time multiplexing period of 20 of the proposed system. Although the AM06 does not meet the latency requirements, the number of patterns per chip are already enough to fulfill the CMS track trigger requirements. The PRM06 reconstruction algorithm was developed to show that a single mezzanine can deal with the number of patterns needed in a trigger tower. Indeed, the number of patterns needed to cover a trigger tower with a satisfying track reconstruction efficiency were evaluated to be in a range between 0.5M and 1M depending on the angular (η) region of the trigger tower.

Another specific feature of the PRM06 reconstruction algorithm is the way it filters the stubs that are retrieved from the DO after the pattern matching phase. The filtering phase, after the stub retrieval from the DO, is a necessary phase to clean out spurious stubs from the obtained stub set. This is due to the fact that coarse resolution stub data (SSIDs) are used to match the AM chip patterns and multiple stubs can end up creating the same SSID. Once the stubs related to the SSIDs of a pattern are retrieved from the DO, more than one stub per layer might be found. Since the track fitter algorithm accepts only track candidates with a single stub per layer, a filter stage is implemented with a pattern finding algorithm, which uses combinations of pairs of stubs coming from the first three layers of the outer tracker as seeds to find the other track-compatible stubs. The filtering module outputs only the best combination of stubs, fulfilling the requirements of a single stub per layer and only one possible missing stub in one layer.

The track reconstruction algorithm of the PRM06 board was checked by comparing the output of the test stand with the HDL simulation and with the C++ emulation of the system. Various simulated event samples were used to test the board and matching AM06 pattern recognition output was obtained when comparing the emulation and HDL simulation, demonstrating the correct performance of the PRM06 board.

The FNAL PRM reconstruction algorithm was intended to measure the latency of the track reconstruction algorithm, emulating the operational behavior of the future AM chip and testing the algorithms with the current available FPGAs. To this end, all track reconstruction steps shown in Figure 4 were optimized to reduce the latency to a minimum and to foster pipelines in the system.

Apart from the different goals of the FNAL PRM and PRM06, the main differences in the track reconstruction algorithm are the different ways of dealing with the spurious stubs after the stub retrieval phase from the DO and the track fitting method precision. The FNAL PRM algorithm does not filter the spurious stubs out, instead implementing an algorithm that generates all possible permutations of one stub per layer and all possible permutations of one stub per layer with one stub missing in one layer. This method has low latency when there are only a few stubs per layer, but it is suboptimal in case of large number of stubs per layer: in this case, the number of permutations can be very large, increasing the latency of the system. To prevent the number of permutations from affecting the overall performance of the system, a truncation on the number of permuations considered is set to 480. The truncation limit has never been met with any simulated event samples considered (even in the worst case: $t\bar{t}$ events with pileup of 250). In both FNAL PRM and PRM06, a linearized χ^2 track fitter (PCA) was implemented with HDL code in the FPGA. The FNAL PRM track fitter was designed to give more precise parameters by adding higher-order corrections to the linearized fitting matrices.

6 Demonstrator performance

Using simulated event samples, we tested the prototype by splitting the total time of processing and latencies into three main logical blocks. The first logical block is the data delivery latency. It is the time that is needed to bring the stubs of an event from the outer tracker front-end electronics to the mezzanine interface of the Pulsar IIb. It was measured in hardware using the two-ATCA shelf prototype and it also accounts for the latency needed for the data formatting and data delivery (serialization and deserialization latency is included). The second logical block is the pattern recognition latency; it includes the latency due to the AM chip pattern recognition, the latency due to the retrieving of pattern SSIDs from the memory, and the retrieving of the stubs per each SSID from the DO. The third logical block is the track fitting latency; it includes the latency of the permutation builder and the latency of the PCA track fitter.

For the latency and total time of reconstruction measurements we used the FNAL mezzanine that was developed for this purpose. All the HDL algorithms implemented in the FPGA were running at the frequency of 240 MHz. The FPGA emulating the AM chips can contain only about 4k patterns, but they were enough to test single events, since the number of matching patterns is one order of magnitude less. In the test configuration, the FNAL mezzanine could serve one fourth of the trigger tower, since there was only one emulated AM chip, while in the future mezzanine, at least four chips are expected. To optimize the pipelining of the algorithms three instances of permutation builder and PCA track fitter were implemented.

The three logical blocks in which the latency is split are fully pipelined. Thus, when the first data exits one block it is immediately computed by the next block. In Table 1, we report the time measurements performed with the demonstrator. The table reports the starting and ending time for each logical block. The total time for the track reconstruction is estimated to be 2.53 μs.

The track parameter performance of the system is generally good. Muons are reconstructed with an efficiency greater than 95% for p_T greater than 3 GeV/c. Good performance is obtained also for pions. The track parameter resolutions are excellent; it is worth mentioning that the p_T resolution over p_T of the muons is at the order of 1%.

Table 1. Measured time for each logical block of the demonstrator.

Logical block	Starting time (μs)	Ending time (μs)
Data delivery to the PRM	1.2	1.7
Pattern recognition	1.85	2.3
Track fitting	2.03	2.53

7 Conclusions

A low latency track reconstruction demonstrator for the L1 trigger of CMS for the HL phase of the LHC has been built. In this approach, we made use of custom AM chips and commercial FPGAs. The demonstrator was composed of two ATCA shelves in order to demonstrate that large amounts of data can be delivered to the pattern reconstruction mezzanines. Two PRMs were tested. One shows that with the available custom AM chips, we can build a working mezzanine that can deal with the number of patterns needed for a trigger tower. The other mezzanine was used to emulate the future AM chip features in order to measure the latency of the pattern recognition and track fitting stages. It was demonstrated that extrapolating the current technology to a reasonably more performant technology, the system can reconstruct the tracks in less than 4 μs with very good track parameter resolution and efficiency.

Acknowledgments

This work is partially funded by MIUR Italy, under contract 2012Z23ERZ of PRIN 2012 H-TEAM and the People Program (Marie Curie Actions) of the European Union's Seventh Framework Program FP7/2007-2013/ under REA grant agreement no. 317446 INFIERI "INtelligent Fast Interconnected and Efficient Devices for Frontier Exploitation in Research and Industry".

References

[1] G. Hall [CMS TMTT], Nucl. Instrum. Meth. A **824** (2016) 292. doi:10.1016/j.nima.2015.09.075

[2] L. Skinnari [CMS Collaboration], EPJ Web Conf. **127** (2016) 00017. doi:10.1051/epjconf/201612700017

[3] F. Palla *et al.* [CMS Collaboration], JINST **11** (2016) no.03, C03011. doi:10.1088/1748-0221/11/03/C03011

[4] M. Dell'Orso and L. Ristori, Nucl. Instrum. Meth. A **278** (1989) 436. doi:10.1016/0168-9002(89)90862-0

[5] J. Adelman *et al.* [CDF Collaboration], "The Silicon Vertex Trigger upgrade at CDF," Nucl. Instrum. Meth. A **572** (2007) 361.

[6] F. Alberti *et al.*, JINST **8** (2013) C01040. doi:10.1088/1748-0221/8/01/C01040

[7] C. Gentsos, F. Crescioli, P. Giannetti, D. Magalotti and S. Nikolaidis, "Future evolution of the Fast TracKer (FTK) processing unit," PoS TIPP **2014** (2014) 209.

[8] J. Olsen, T. Liu and Y. Okumura, JINST **9** (2014), C01041. doi:10.1088/1748-0221/9/01/C01041

[9] T. Liu, G. Deptuch, J. Hoff, S. Jindariani, S. Joshi, J. Olsen, N. Tran and M. Trimpl, JINST **10** (2015), C02029. doi:10.1088/1748-0221/10/02/C02029

[10] M. Jeitler *et al.*, JINST **2** (2007) P01006. doi:10.1088/1748-0221/2/01/P01006

[11] A. Andreani, A. Annovi, R. Beccherle, M. Beretta, N. Biesuz, W. Billereau, R. Cipriani, S. Citraro *et al.*, "The Associative Memory Serial Link Processor for the Fast TracKer (FTK) at ATLAS," JINST **9** (2014) 11, C11006.

Track reconstruction for the *Mu3e* experiment based on a novel Multiple Scattering fit

Alexandr **Kozlinskiy**[1,a] for the *Mu3e* Collaboration

[1]*Institut für Kernphysik, Johannes Gutenberg-Universität Mainz, Johann-Joachim-Becher-Weg 45, 55128 Mainz, Germany*

Abstract. The *Mu3e* experiment is designed to search for the lepton flavour violating decay $\mu^+ \to e^+e^+e^-$. The aim of the experiment is to reach a branching ratio sensitivity of 10^{-16}. In a first phase the experiment will be performed at an existing beam line at the Paul-Scherrer Institute (Switzerland) providing 10^8 muons per second, which will allow to reach a sensitivity of $2 \cdot 10^{-15}$. The muons with a momentum of about 28 MeV/c are stopped and decay at rest on a target. The decay products (positrons and electrons) with energies below 53 MeV are measured by a tracking detector consisting of two double layers of 50 µm thin silicon pixel sensors. The high granularity of the pixel detector with a pixel size of 80 µm × 80 µm allows for a precise track reconstruction in the high multiplicity environment of the *Mu3e* experiment, reaching 100 tracks per reconstruction frame of 50 ns in the final phase of the experiment. To deal with such high rates and combinatorics, the *Mu3e* track reconstruction uses a novel fit algorithm that in the simplest case takes into account only the multiple scattering, which allows for a fast online tracking on a GPU based filter farm. An implementation of the 3-dimensional multiple scattering fit based on hit triplets is described. The extension of the fit that takes into account energy losses and pixel size is used for offline track reconstruction. The algorithm and performance of the offline track reconstruction based on a full Geant4 simulation of the *Mu3e* detector are presented.

1 Introduction

The *Mu3e* experiment [2] is designed to search for Lepton Flavor Violation (LFV) in the decay $\mu^+ \to e^+e^+e^-$. The goal of the experiment is to observe this decay or set a limit on the branching fraction at the level of 10^{-16}. To reach such sensitivity, more than 10^{16} muon decays have to be observed and any backgrounds have to be suppressed to the 10^{-16} level. To achieve this goal within one year of data taking will require a muon stopping rate of more than 10^9 Hz with a high geometrical acceptance and efficiency of the detector.

In the Standard Mode (SM) lepton flavor is conserved at tree level. However, flavor is known to be not conserved in the neutrino sector where it is manifested in the form of neutrino oscillations. Contributions of neutrino mixing in loops however lead to an unobservable

[a]e-mail: akozlins@uni-mainz.de

branching fraction for $\mu^+ \to e^+e^+e^-$ of the order of 10^{-54} [1]. Consequently, the study of this decay is of large interest, as the observation of a signal will be a clear sign of New Physics (NP) beyond the Standard Model.

The current limit on the branching fraction of $\mu^+ \to e^+e^+e^-$ comes from the SINDRUM experiment and lies at $B(\mu^+ \to e^+e^+e^-) < 1.0 \cdot 10^{-12}$ at 90% CL [3]. *Mu3e* plans to improve this limit by four orders of magnitude.

The *Mu3e* experiment is planned to be performed in two phases. In the first phase the existing PiE5 beam line at the Paul-Scherrer Institute (PSI) will be used. During this phase the experimental concept and instrumentation will be validated. The goal for the first phase is to reach a sensitivity of $2 \cdot 10^{-15}$, improving the current limit by almost three orders of magnitude. For the second phase the installation of a new high intensity muon beam line (*HiMB*) is required, which has to provide a continuous beam of more than 10^9 muons per second.

1.1 Signal and background

The signal decay $\mu^+ \to e^+e^+e^-$ contains in the final state three particles: two positrons and one electron. The total momentum and energy of these three particles will correspond to the muon at rest ($\sum \boldsymbol{p} = 0$ and $\sum E = m_\mu$) as the muon stops and decays on target. In addition to the energy/momentum signature, the products of decay will have a common vertex and have to coincide in time.

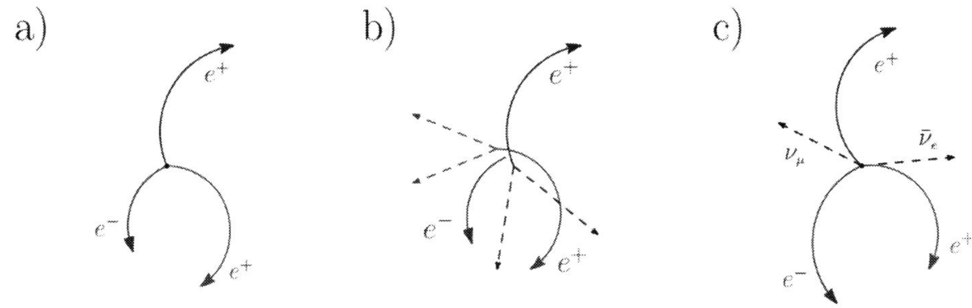

Figure 1. Schematic view of a) signal decay $\mu^+ \to e^+e^+e^-$, b) overlap of several Michel decays, c) internal conversion decay $\mu^+ \to e^+e^+e^-\nu\bar{\nu}$ where two neutrinos carry only part of the energy.

There are two sources of background. One is the combinatorial background due to several processes overlapping such that they mimic a signal decay, i.e. a positron from a Michel decay ($\mu^+ \to e^+\nu\bar{\nu}$) and an electron/positron pair from Bhabha scattering. Figure 1b shows a schematic view of such background. It can be suppressed with vertexing, good momentum and time resolution.

The second source is the internal conversion decay $\mu^+ \to e^+e^+e^-\nu\bar{\nu}$ where the only difference from the signal decay is missing energy and momentum. This type of background can only be suppressed with excellent momentum and energy resolution. Figure 2 shows the branching fraction of the internal conversion decay depending on the cut on missing energy [6]. At the level of 10^{-16}, the separation of signal and background is about 1.2 MeV, which requires a single particle momentum resolution of better than 0.5 MeV/c.

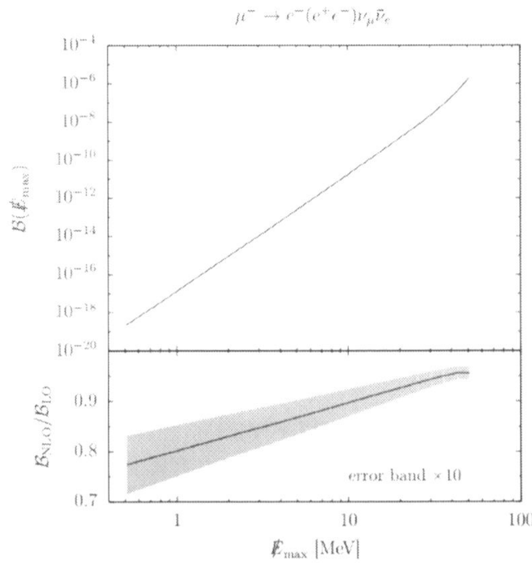

Figure 2. Branching fraction of decay $\mu^+ \to e^+e^+e^-\nu\bar{\nu}$ as function of cut on missing energy $\not{E} = m_\mu - E_{tot} < \not{E}_{max}$ [6]. At 10^{-16} level the cut lies at $\not{E}_{max} \approx 1.2$ MeV.

2 The *Mu3e* detector

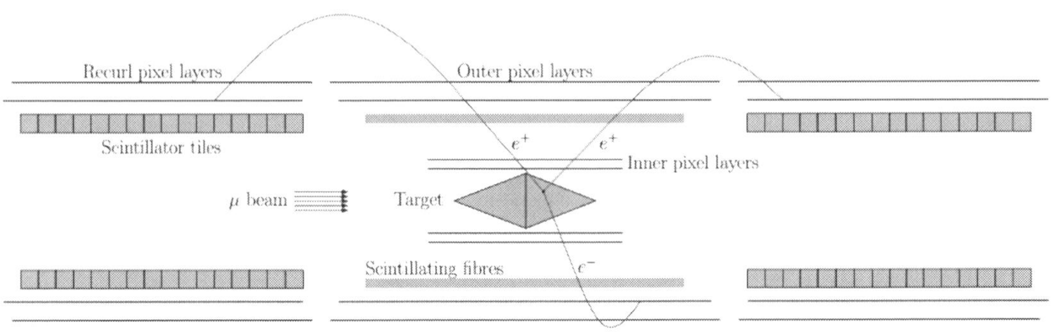

Figure 3. Schematic view of *Mu3e* detector cut open along beam line in the Phase I configuration. Not to scale.

The *Mu3e* detector has a design that is optimized for the detection of the $\mu^+ \to e^+e^+e^-$ signal decay. The whole detector is placed inside a 1 T solenoid magnet with 1 m bore and 2.5 m length. Figure 3 shows a schematic view of the phase I detector.

The muons with a momentum of about 28 MeV/c are stopped on a double cone hollow target and decay at rest. The target is made from 75 (85) µm of Mylar in the front (back) part and has a radius of 19 mm and length 100 mm which allows to spread vertices in space, reducing background due to random combinations of tracks and in addition helping to lower the occupancy in the tracking detector. The products of the decays are detected by two double layers of silicon pixel sensors placed around the target. As the muon decays at rest, the maximum energy of the electrons/positrons produced is half the muon mass or about 53 MeV. Such low energies require to reduce as much as possible the amount of material seen by the particle, as multiple Coulomb scattering dominates the tracking uncertainty.

Mu3e will use High-Voltage Monolithic Active Pixel sensors (HV-MAPS [4]) that are based on a commercial HV-CMOS technology. These sensors can be thinned down to 50 µm which is less than 10^{-3} of a radiation length. All sensors of the tracker will have a size of 20×20 mm^2 with a pixel size of 80×80 µm^2. The sensors are mounted on a 25 µm kapton frame that provides mechanical support and a 50 µm Kapton/Aluminum readout flex print is used for power and signal routing.

The two inner layers consist of 8 and 10 ladders of six sensors and are placed as close as possible to the target to reduce the effect of multiple scattering and pixel size on the precision of the reconstructed vertex. The inner layers have a distance of about 23 mm and 29 mm respectively from the beam axis. The outer layers consist of 24 and 28 ladders of 18 sensors (total length of 36 cm) and have a distance to the beam axis of about 70 mm and 82 mm. The placement of the outer layer sets the limit on the minimum transverse momentum acceptance which is about 12 MeV/c.

Due to their low energy it is possible to measure particles when they bend back in the magnetic field. Such recurling particles will have better momentum resolution due to the constraint on the measured curvature (momentum) being limited only by the pixel size. To increase the acceptance for such recurling particles, two additional stations of double silicon layers are placed upstream and downstream of the central detector. The recurl stations have the same geometry as the outer double layers of the central station. The total length of the detector with recurl stations is about 1 m.

To further suppress accidental background, two timing detectors are used. Scintillating fibres are placed before the outer double layer of the central station and scintillating tiles are placed inside the recurl stations. The fibres of 28 cm length are arranged in ribbons that consist of three layers of 250 µm fibres and provide a time measurement with $\sigma_t \approx 0.5$ ns. The much thicker tiles will provide a time measurement for recurling tracks with $\sigma_t \approx 100$ ps.

The hit information will be readout continuously and assembled into 50 ns reconstruction frames. The data from all parts of the detector are collected by a readout system organized as a switching network consisting of FPGA boards and Gbit/s optical links. The combined data streams are routed to a filter farm where tracks and vertices are reconstructed.

3 Reconstruction

The reconstruction performance is studied on Monte Carlo (MC) data obtained with a full Geant4 simulation of the *Mu3e* detector. The simulated data are digitized and processed through the readout chain, where they are sorted and assembled to frames similar to real data. At 50 ns frame size and a muon stopping rate of $2 \cdot 10^8$ Hz there are about five Michel decays in each frame within the acceptance of the detector.

Each Michel positron produces on average two hits in each layer of the detector: one on the way out of the detector volume and another one when it recurls back. To identify tracks, it is necessary to find hits produced by the same particle and assemble them in the right order. In case of *Mu3e*, due to the low momentum, the hits from the same particle do not form straight segments and can be separated by large distances. This makes the identification of tracks difficult and requires to consider many possible combinations of hits. As a minimum of three hits are necessary to define an initial trajectory or seed, the number of combinations that have to be processed or fitted scales as the third power of number of hits in a layer. For $n = 10$ the number of triplet candidate can reach 1000 per frame, which requires to be able to fit up to 10^{10} triplet candidates each second.

As it is not possible to store all the data produced by the detector, the online filter farm will be used to reduce the data rate by performing a full reconstruction of each frame [5]. To cope with such high rates and be able to fully reconstructed each frame *Mu3e* will use the novel 3-dimensional Multiple Scattering (MS) triplet fit [7].

3.1 Tripet fit

The triplet fit is designed for environments where the hit position uncertainties can be neglected compared to the multiple scattering. This is the case for the *Mu3e* experiment, where due to the low energies of particles multiple scattering dominates over other effects, such as pixel size.

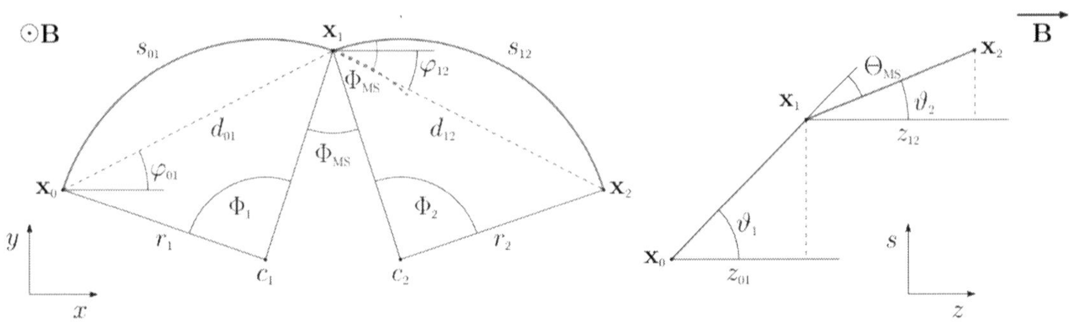

Figure 4. Schematic view of a triplet in the transverse (left) and longitudinal (right) plane. The green line shows a particle trajectory consisting of two helices between the hits x_0 and x_1, and x_1 and x_2. The scattering is defined at the middle hit x_1 with projected MS angles φ_{MS} and θ_{MS}.

The idea of the triplet fit comes from the fact that in the magnetic field the trajectory of the particle can be fully described by knowing the position of three hits (in 3D space) and the curvature (momentum), assuming negligible hit uncertainties and energy losses. In this scenario the trajectory consists of two helices between the first and second pair of hits and a kink at the middle hit, attributed to scattering, as shown in figure 4. The parameters of both helices are completely defined by the 3D curvature r of the trajectory. Consequently the scattering angle projections φ_{MS} and θ_{MS} are functions of r.

The fit can be defined as the trajectory that minimizes the total scattering angle or χ^2

$$\chi^2(r) = \varphi^2_{MS}(r)/\sin^2(\theta) + \theta^2_{MS}(r), \tag{1}$$

where θ is the helix polar angle.

The minimization problem is non-linear and does not have an analytic solution. However, the problem can be linearized around some known solution such that both φ_{MS} and θ_{MS} are linear functions of the curvature r. In case of small scattering angles, the initial solution can be selected as the trajectory that is circular in the transverse xy-plane. For this trajectory, the scattering angle in the transverse plane is zero and parameters of both helices can be calculated from the positions of the hits. The solution to the minimization of the χ^2 in this linear case can be calculated [7].

In the case of tracks with more than three hits, the trajectory can be divided into a sequence of triplets such that each pair of triplets shares two hits. Due to the quadratic form of the triplet χ^2, it can be shown that the solution to the minimization of the track χ^2 is a weighted average of individual triplet solutions

$$r_{track} = \frac{\sum r_i \sigma_{r,i}^{-2}}{\sum \sigma_{r,i}^{-2}}, \tag{2}$$

where r_i and $\sigma_{r,i}$ are the curvature and uncertainty for triplet i.

The triplets can be fitted separately and subsequently combined together, however in practice the reconstruction starts with a seed triplet and additional hits are then added to the seed forming new triplets.

3.2 Track reconstruction

The reconstruction is performed in several steps. In a first step, the triplet seeds are formed which are subsequently used in the reconstruction of tracks. To reduce the number of triplet combinations, geometrical cuts on distances and opening angles between hits are used. After this, the triplets are fitted and a loose χ^2 cut is used to further reduce the number of wrong (fake) hit combinations. Still, the set of reconstructed triplets contains a large number of fakes, reaching 4 wrong for each truth matched combination.

Each triplet is then used as a seed for finding and reconstructing tracks that consist of at least 4 hits, with at least one hit in each layer. These are called short tracks and they correspond to the out- and in-going parts of the particle trajectory. For each triplet, the trajectory is extended to the outer layer and if there is a hit in the vicinity of the intersection of the layer with the trajectory, this hit is added to the triplet to form a short track. After the reconstruction of short tracks, the fake rate is reduced to the percent level. Figure 5a shows reconstructed triplets and short tracks in the transverse view of one reconstructed frame.

The short tracks are used for the reconstruction of recurling particles that form long tracks. There are two types of such tracks: 8- and 6- hits tracks. Long 8-hit tracks correspond to one full turn in the central part of the detector. Such tracks are reconstructed by combining two short tracks. Long 6-hit tracks are reconstructed from a short track and a pair of hits in two outer layers, either in the central or recurl stations. Figure 5b illustrates reconstructed long tracks in the transverse view of one reconstructed frame.

The fake rate for long 6-hit tracks is about 5%. For long 8-hit tracks the number of fakes can reach the number of truth tracks. Such a high fake rate is due to particles that make several turns in the magnetic field in the central part of the detector volume where wrong combinations are made from two short tracks that correspond to different turns.

For all types of track the fake rate due to the combination of hits from different tracks is of the order of 1%. The rest is coming from combinations of hits from the same track but different turns and can be reduced by applying cuts on the inclination angle of tracks, rejecting so-called vertical tracks.

In addition to random combinations, noise hits can contribute to the fake rate. Sensor noise is expected at a level of 0.1 Hz per pixel [8] which corresponds to about one noise hit per frame of 50 ns. At noise levels up to 10 Hz per pixel the contribution to the fake rate and the reconstruction efficiency is negligible.

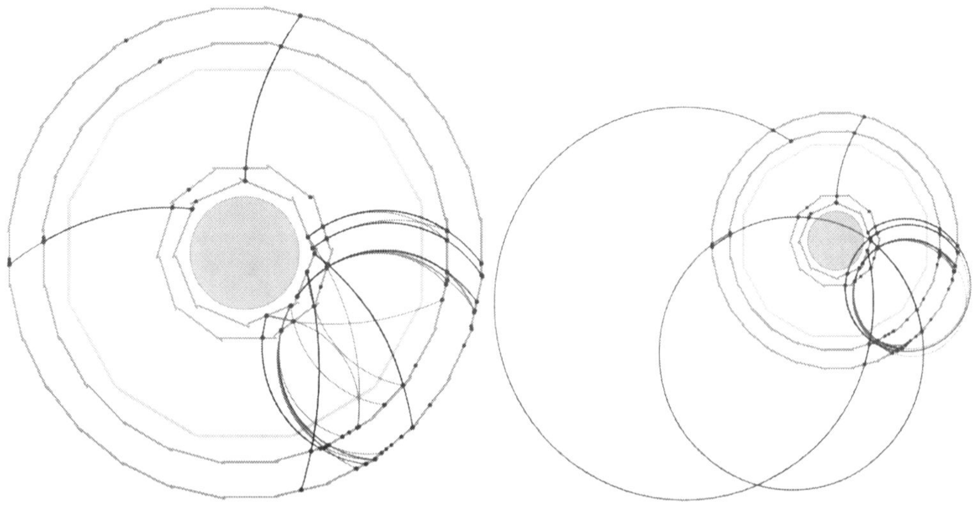

Figure 5. Transverse view of a reconstructed frame showing a) triplets and shorts tracks, b) long tracks. Truth matched and fake track candidates are shown in black and red respectively.

3.3 Performance

Figure 6 shows the total reconstruction efficiency (including acceptance) as a function of momentum p and inclination angle $\lambda = \pi/2 - \theta$.

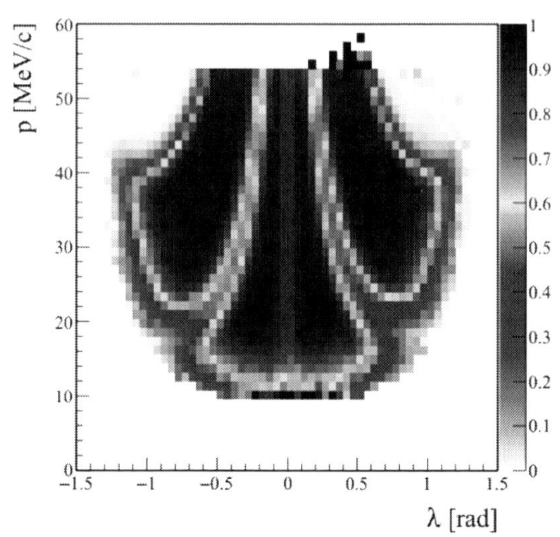

Figure 6. Total reconstruction efficiency for long tracks as a function of momentum p and inclination angle $\lambda = \pi/2 - \theta$. The edges of the distribution are defined by the geometry of the detector (radius of outer layer, length of the detector and gaps between stations).

The acceptance of the detector is limited by the outer radius of the detector which is represented by the lower edge in the efficiency plot. The missing corners are due to the limited length of the detector and vertical stripes correspond to service areas between the central and recurl stations. The efficiency is limited by geometrical cuts that are used to make the initial triplet seeds and by track χ^2 cuts.

The achieved momentum resolution for short and long tracks is shown in figure 7. The average momentum resolution for short tracks is $\sigma_p = 1.4$ MeV/c and for long tracks it is $\sigma_p = 0.2$ MeV/c.

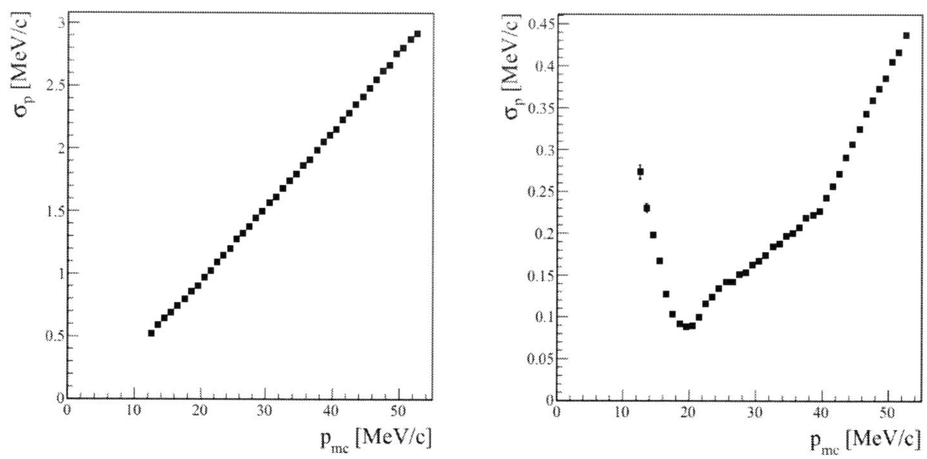

Figure 7. Momentum resolution for short (left) and long (right) tracks. The average momentum resolution is 1.4 MeV/c and 0.2 MeV/c respectively.

3.4 Charge ID

The timing information is provided by the fibre and tile detectors and is essential to reduce background due to random combinations of tracks. Each short track crosses the fibre detector and is assigned a fibre cluster. The long 8-hit tracks can have 2 fibre clusters, one for each short track.

The timing information is also used to resolve the charge ambiguity for reconstructed 8-hit long tracks. This ambiguity arises due to the unknown direction of motion of recurling particles. The fibre tracker allows to resolve the ambiguity by looking at the time difference between fibre hits. Figure 8 shows the time difference between two fibre hits as function of the corresponding path length. The positive time difference corresponds to right direction (charge) and negative time corresponds to the wrong charge id. The slope of the distribution corresponds to the speed of light.

3.5 Vertex fit and signal sensitivity

The final step of the reconstruction is the vertex fit. Three tracks from the signal decay should originate form a common point on the surface of the target. During the vertex fit, the

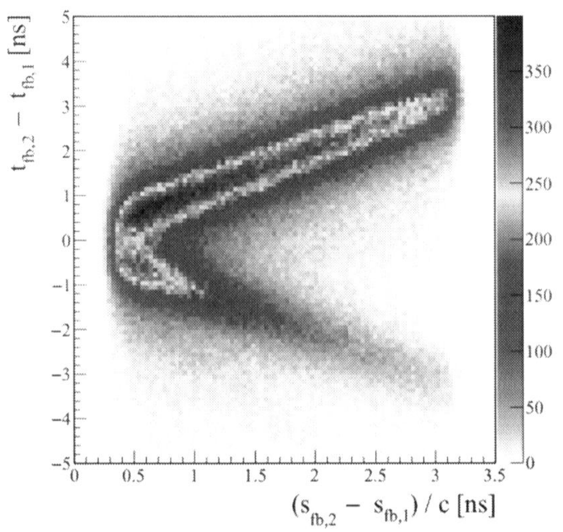

Figure 8. Distribution of the time difference between two fibre hits of recurling particle as a function of path length along the reconstructed trajectory from the first to the second fibre hit.

scattering in inner layers and the curvature uncertainties are taken into account. To suppress background, fit χ^2 and distance to target can be used. Figure 9 shows the distribution of distances between the reconstructed vertex and truth MC vertex position.

To estimate the signal sensitivity of the *Mu3e* experiment, simulated events with one signal decay in 50 ns reconstruction frame were generated. All combinations of two positively and one negatively charged tracks are tried to form a vertex that is fitted in the inner region of the detector. After the vertex fit, the invariant mass of the candidate signal decay is estimated from the parameters of the tracks at the vertex position. The reconstructed mass of candidates is shown in figure 9.

4 Summary

The *Mu3e* experiment is designed to search for the LFV decay $\mu^+ \to e^+e^+e^-$. The detector performance was studied with the full Geant4 detector simulation and reconstruction. In one year (10^7 seconds) of data taking with the muon stopping rate of $2 \cdot 10^8$ Hz on target *Mu3e* will be able to reach a single event sensitivity at the $2 \cdot 10^{-15}$ level. This corresponds to the target sensitivity of the first phase of the experiment. The current detector design is shown to reach and exceed the sensitivity requirements in terms of momentum and invariant mass resolution.

References

[1] W. J. Marciano, T. Mori and J. M. Roney, "Charged Lepton Flavor Violation Experiments," Ann. Rev. Nucl. Part. Sci. **58** (2008) 315. doi:10.1146/annurev.nucl.58.110707.171126

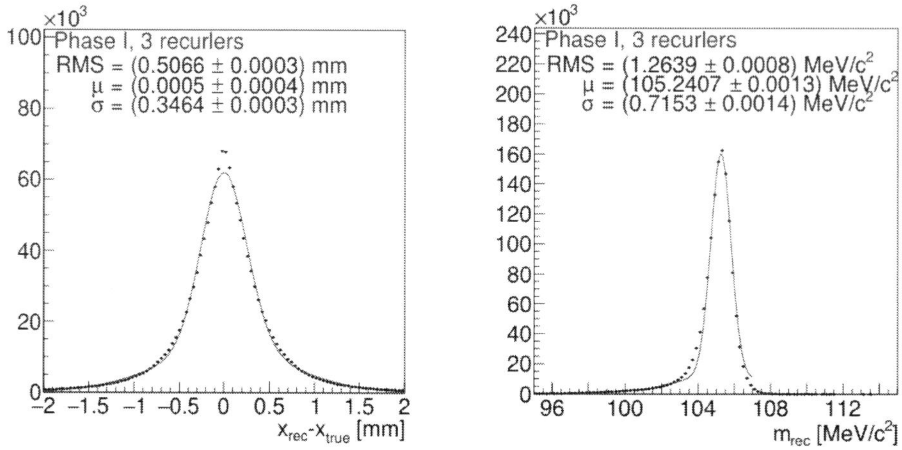

Figure 9. Distance from truth vertex position x_{true} to the reconstructed vertex x_{rec} (left) and invariant mass distribution of truth candidates.

[2] A. Blondel, A. Bravar, M. Pohl, S. Bachmann, N. Berger, M. Kiehn, A. Schoning and D. Wiedner *et al.*, "Research Proposal for an Experiment to Search for the Decay $\mu^+ \to e^+e^+e^-$," arXiv:1301.6113 [physics.ins-det].

[3] U. Bellgardt *et al.* [SINDRUM Collaboration], "Search for the Decay $\mu^+ \to e^+e^+e^-$," Nucl. Phys. B **299**, 1 (1988).

[4] I. Peric, "A novel monolithic pixelated particle detector implemented in high-voltage CMOS technology," Nucl. Instrum. Meth. A **582**, 876 (2007).

[5] D. vom Bruch, "Online Track and Vertex Reconstruction on GPUs for the Mu3e Experiment," Workshop "Connecting The Dots / Intelligent Trackers 2017", LAL-Orsay.

[6] M. Fael and C. Greub, "Next-to-leading order prediction for the decay $\mu \to e\ (e^+e^-)\ \nu\,\overline{\nu}$," JHEP **1701** (2017) 084 doi:10.1007/JHEP01(2017)084 [arXiv:1611.03726 [hep-ph]].

[7] N. Berger, M. Kiehn, A. Kozlinskiy and A. Schöning, "A New Three-Dimensional Track Fit with Multiple Scattering," Nucl. Instrum. Meth. C **844** (2017) 135 doi:10.1016/j.nima.2016.11.012 [arXiv:1606.04990 [physics.ins-det]].

[8] H. Augustin *et al.*, JINST **11** (2016) no.11, C11029 doi:10.1088/1748-0221/11/11/C11029 [arXiv:1610.02210 [physics.ins-det]].

Parallelized Kalman-Filter-Based Reconstruction of Particle Tracks on Many-Core Processors and GPUs

Giuseppe **Cerati**[4,a], Peter **Elmer**[2,b], Slava **Krutelyov**[1,c], Steven **Lantz**[3,d], Matthieu **Lefebvre**[2,e], Mario **Masciovecchio**[1,f], Kevin **McDermott**[3,g], Daniel **Riley**[3,h], Matevž **Tadel**[1,i], Peter **Wittich**[3,j], Frank **Würthwein**[1,k], and Avi **Yagil**[1,l]

[1]*UC San Diego, La Jolla, CA, United States of America 92093*
[2]*Princeton University, Princeton, NJ, United States of America 08544*
[3]*Cornell University, Ithaca, NY, United States of America 14853*
[4]*Fermilab, Batavia, IL, United States of America 60510-5011*

Abstract. For over a decade now, physical and energy constraints have limited clock speed improvements in commodity microprocessors. Instead, chipmakers have been pushed into producing lower-power, multi-core processors such as Graphical Processing Units (GPU), ARM CPUs, and Intel MICs. Broad-based efforts from manufacturers and developers have been devoted to making these processors user-friendly enough to perform general computations. However, extracting performance from a larger number of cores, as well as specialized vector or SIMD units, requires special care in algorithm design and code optimization. One of the most computationally challenging problems in high-energy particle experiments is finding and fitting the charged-particle tracks during event reconstruction. This is expected to become by far the dominant problem at the High-Luminosity Large Hadron Collider (HL-LHC), for example. Today the most common track finding methods are those based on the Kalman filter. Experience with Kalman techniques on real tracking detector systems has shown that they are robust and provide high physics performance. This is why they are currently in use at the LHC, both in the trigger and offline. Previously we reported on the significant parallel speedups that resulted from our investigations to adapt Kalman filters to track fitting and track building on Intel Xeon and Xeon Phi. Here, we discuss our progresses toward the understanding of these processors and the new developments to port the Kalman filter to NVIDIA GPUs.

[a]e-mail: giuseppe.cerati@cern.ch
[b]e-mail: peter.elmer@cern.ch
[c]e-mail: vyacheslav.krutelyov@cern.ch
[d]e-mail: steve.lantz@cornell.edu
[e]e-mail: ml15@princeton.edu
[f]e-mail: mario.masciovecchio@cern.ch
[g]e-mail: kevin.mcdermott@cern.ch
[h]e-mail: daniel.riley@cornell.edu
[i]e-mail: matevz.tadel@cern.ch
[j]e-mail: wittich@cornell.edu
[k]e-mail: fkw@ucsd.edu
[l]e-mail: ayagil@physics.ucsd.edu

1 Introduction

The Large Hadron Collider (LHC) at CERN is the highest energy collider ever constructed. It accelerates two counter-circulating proton beams and brings them to collision in four locations around a 27 kilometer ring straddling the border between Switzerland and France. By several measures it is the largest man-made scientific device on the planet. The goal of the LHC is to probe the basic building blocks of matter and their interactions. In 2012, the Higgs boson was discovered by the CMS and ATLAS collaborations [1, 2]. By measuring the energy and momentum of particles produced by the collision, we can infer the existence of massive particles that were created and measure their properties. This process is known as event reconstruction and consists of integrating information from different detector components. Track reconstruction, also known as tracking, is one step in event reconstruction and determines the trajectories of charged particles ("tracks") from a set of positions of energy deposits within the various layers of our detectors ("hits"). Tracking, in comparison to other aspects of event reconstruction, is the most computationally time consuming, the most sensitive to increased activity in the detector, and traditionally, the least amenable to parallelized processing. The speed of online reconstruction has a direct impact on how much data can be stored from the 40 MHz collisions rate, while the speed on the offline reconstruction limits how much data can be processed for physics analyses. This research is aimed at vastly speeding up tracking.

The large amount of time spent in tracking will become even more important in the high-luminosity era of the LHC. The increase in event rate will lead to an increase in detector occupancy ("pile-up", PU), leading to an exponential gain in the time required to perform track reconstruction, as can be seen in Figure 1 [3]. In the Figure, PU25 corresponds to the data taken during 2012, and PU140 corresponds to the low end of estimates for the HL-LHC era. Clearly our research on tracking performance will become increasingly important during this era.

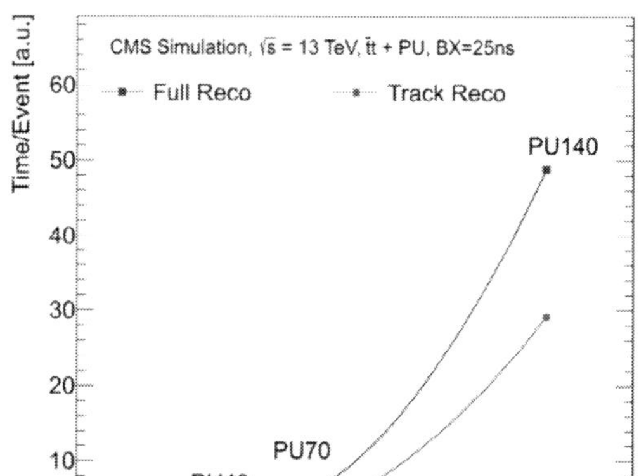

Figure 1: CPU time per event versus instantaneous luminosity, for both full reconstruction and the dominant tracking portion. PU25 corresponds to the data taken during 2012, and PU140 corresponds to the HL-LHC era. The time of the reconstruction is dominated by track reconstruction [3].

Accommodating this increasing need for computational power is complicated by the change in computing architectures during the last decade. Around 2005, the computing processor market reached a turning point: power density limitations in chips ended the long-time trend of ever-increasing clock speeds, and our applications no longer immediately run exponentially faster on subsequent generations of processors. This is true even though the underlying transistor count continues to increase per Moore's law. Increases in processing speed such as those required by the time increases in Figure 1 will no longer come 'for free' from faster processors. New, parallel processors instead are aggregates of 'small cores' that in total still show large performance gains from generation to generation, but their usage requires reworking of our mostly serial software to exploit these gains. The processors in question include ARM, Graphical Processing Units (GPU) and the Intel Xeon and Xeon Phi.

In what follows, we first update the reader about the algorithms we target and the performance level we now obtain on Xeon and Xeon Phi. We then explain the challenges and the steps we took in porting these algorithms to GPUs along with some initial performance assessments.

2 Kalman Filter Tracking

The algorithm we are targeting for parallelized processing is a Kalman Filter (KF) based algorithm [4]. KF-based tracking algorithms are widely used since they are fast and incorporate estimates of multiple scattering in the detector material directly into the trajectory of the particle. In addition, a combinatorial KF provides an elegant way to resolve ambiguities when considering multiple hits to add to a track [5, 6]. Other track finding algorithms, more naturally suited to parallelization and coming from the image processing community, have been explored by others. These include Hough Transforms and Cellular Automata, among others (see, for instance, [7]). However, these are not the main algorithms in use at the LHC today, whereas there is extensive understanding on how KF algorithms perform. KF algorithms have proven to be robust and perform well in the difficult experimental environment of the LHC [3]. Rather than abandon the collective knowledge gained from KF algorithms, we wish to extend this robust tool by porting KF tracking to parallel architectures.

KF tracking proceeds in three main stages: seeding, building, and fitting. Seeding provides the initial estimate of the track parameters based on a few hits in a subset of layers. Realistic seeding is currently under development and will not be reported here. Building then collects additional hits in other detector layers to form a complete track, using the combinatorial KF as a basis for deciding which hits to consider and keep. Track building is by far the most time consuming step of tracking, as it requires branching to explore potentially more than one track candidate per seed after finding compatible hits on a given layer. After hits have been assigned to tracks in each layer, a final fit with a KF and smoother can be performed over each track to provide the best estimate of its parameters and to compute a measure of the track quality in terms of a χ^2.

To realize performance gains, we need to exploit two types of parallelism: vectorization and parallelization. Vectorization aims to perform a single instruction on multiple data at the same time by performing the same operation across different data in lockstep. Parallelization aims to perform different instructions at the same time on different data. In this work, parallelization is done by distributing the workload between threads, which are concurrent lightweight processes sharing resources (for instance, memory). The challenge to vectorization is that KF tracking may branch to explore multiple candidates per seed, interfering with the lockstep synchronization required for vectorization performance. The challenge to parallelization is that hit occupancy in a detector is not uniformly distributed on a per event basis, creating the potential for workload imbalances across threads. For these and other reasons, KF tracking cannot be ported in a straightforward way to run in parallel on many-core processors. Past work by our group has shown progress in porting sub-stages of KF tracking to support parallelism in simplified detectors (see, e.g. our presentations at ACAT2014 [8], CHEP2015 [9], and NSS-MIC2015 [10]). As the hit collection is completely determined after track building, track fitting can repeatedly apply the KF algorithm without branching, making this the ideal place to start in porting KF tracking to Xeon and Xeon Phi, with our first results shown at ACAT2014 [8]. In more recent works (CHEP2016 [11], CtD2016 [12]), we discussed the improvements made to target a more realistic geometry. Profiling and recent computational optimizations were also outlined.

2.1 Optimized Matrix Library Matriplex

The computational problem of KF-based tracking consists of a sequence of matrix operations. Unlike [4], we parameterize the measurement state and the track state in global coordinates with matrices of size $N \times N = 3 \times 3$ and 6×6, respectively. To allow maximum flexibility for exploring SIMD operations involving many small-dimensional matrices, and to decouple the specific matrix computations from the high level algorithm, we have developed a matrix library, Matriplex, which relies on a matrix-major memory layout (a layout where elements with the same index but from different matrices are stored consecutively). A more complete description of the Matriplex library can be found in the work we presented at ACAT2014 [8]. The adaptation of this data structure to GPUs will be discussed in section 4.

3 Track Building on Xeon and Xeon Phi

Combinatorial KF track building shares the same core KF calculations as track fitting, but has two major complications for utilizing vectorizable operations compared to track fitting. The first such problem is the "nHits" problem: when performing the KF operations, the hit set for a given track candidate in building is undefined and one potentially has to choose between $\mathcal{O}(10^4)$ hits per layer. The second problem is the "combinatorial" problem: when more than one hit on a given layer is compatible with a given input candidate, branching occurs to explore all possible track candidates up to some cutoff number of candidates per seed.

The key to reducing the nHits problem is to partition the tracks and hits spatially in order to reduce the search space for a given track. We partition space without reference to the detector structure, placing hits in each layer into directionally projective bins. The bins provide a fast look-up of compatible hits in a layer for a given track given the estimate from the KF.

With regard to the combinatorial problem, we first developed track building to add only the best hit out of all compatible hits on a layer for each track candidate. By definition, then, each seed only produces one track and does not require copying of track candidates to explore multiple hits per layer, as in the full combinatorial KF approach. The best hit is defined as the hit that produced the lowest χ^2 increment to the track candidate. The vectorization and parallelization performance of this "best hit" approach were presented at CHEP2015 [9]. After demonstrating the feasibility of track building in the best hit approach, we then allowed for more than one track candidate per seed per layer to be produced, as in the full combinatorial KF approach and presented the results at CtD2016 [12].

3.1 Memory Management

With the full combinatorial approach in place, we performed extensive studies of the performance of our software, in terms of both the physics performance and the code performance. For the latter, we used the Intel VTune 2016 [13] suite of tools to identify bottlenecks and understand the effects of our optimization attempts. In particular, as can be expected, we determined that memory management is of critical importance.

In the full combinatorial approach, if multiple compatible hits are found for a track candidate, it becomes necessary to make copies of that track for each compatible hit. To mitigate the impact from this serial work, we moved copying outside of all vectorizable operations into what we term the "clone engine". The clone engine approach only copies the best N track candidates per seed after reading and sorting a bookkeeping list of all compatible hits. This is in contrast to a first attempt at the combinatorial KF which copied a candidate each time a hit was deemed compatible, then sorting and keeping only the best N candidates per seed after all the possible hits on a given layer for all input candidates were explored. A more detailed discussion of this work on memory management and impacts on performance was presented at NSS-MIC2015 [10] and at CHEP 2016 [11]

3.2 Results

We present here the latest vectorization and parallelization benchmarks in track building since CHEP 2016. Results are given in average time per event. For each event, building proceeds from $10,000$ seeds that are derived from the first three hits of each simulated track in the event. We used a simplified detector geometry consisting of equally spaced, fixed length concentric cylinders. Tracks are simulated using a set of in-house simulation routines replicating the detector resolution via Gaussian smearing of the hit positions. Figure 2 contains two plots displaying the performance obtained on a computer with two Sandy Bridge multi-core processors (E5-2620 @ 2.00GHz). Plot 2a shows the vectorization performance of three different track building approaches, as functions of the number of vector units enabled while using a single thread. Plot 2b shows the parallelization performance of the same track building approaches, as functions of the number of threads enabled using Intel Thread Building Blocks (TBB) [14]. The Xeon machine we are using has 12 physical cores which appear to be 24 logical cores to the OS due to hyperthreading being enabled. The slight deviation from ideal scaling before hyperthreading arises from load balancing issues between threads. The large deviation from ideal scaling after enabling hyperthreading is due to resource limitations: the two threads per core are contending for the same instruction pipelines and data caches. Even so, nearly eightfold speedup is seen for 21 threads.

The blue curve is the "best hit" approach described previously. Naturally, this approach will have the lowest absolute time in comparison to the two fully combinatorial approaches, the red and green curves. The green curve is the original approach to combinatorial track finding where the copying of the track candidates is inside the vectorizable KF operations. The red curve shows the clone engine approach, which moves the serial work of copying outside of the KF vectorizable operations. As expected, the clone engine approach has a lower absolute time and higher speedup than the original. It is important to note that while the best hit approach is the fastest, the physics performance has inefficiencies in hit finding and track finding from not being fully able to explore multiple track candidates per seed. Even in our simplified model, this behavior is already apparent, and the best hit approach is expected to become even more inefficient when using more realistic detector effects. For reference, on this simplified problem, the best hit approach approximately has as an efficiency of 93% and a fake rate of 3%, while the combinatorial approaches have efficiencies over 99% and fake rates under 1%. (Candidates matching fewer than 7 hits of a simulated track count as neither successes nor fakes.)

Figure 3 shows the same plots as Figure 2, now with Xeon Phi (Knights Corner 7120P). Figure 3a shows the vectorization results with Xeon Phi which has AVX-512 vector registers that are twice as wide as those in Xeon. With one thread enabled, Xeon Phi sees the same $1.5\times$ to $2\times$ speedup as Xeon. As seen on Xeon, the best hit approach still has the lowest absolute timing. Figure 3b displays the parallelization performance, with Xeon Phi, with the full vector width of 16 floats enabled using TBB. The Xeon Phi we are using has 61 physical cores requiring 122 independent instruction streams for full utilization, due to the fact that the Xeon Phi issues instructions for a given thread every other

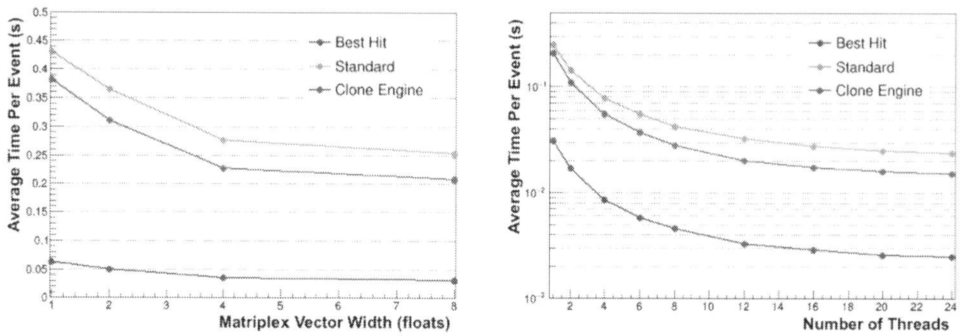

(a) Impact of vector size on performance. Only one thread is enabled.

(b) Impact of the number of threads on performance. The full vector size (8 floats) is enabled.

Figure 2: Average time per event of 10,000 tracks on Xeon (Sandy Bridge E5-2620) processors. Both plots show results for three different track building approaches. The blue curve is the "best hit" tracking approach. The green and red curves use the full combinatorial approach in track building. The red curve moves the copying of tracks outside of the vectorizable operations.

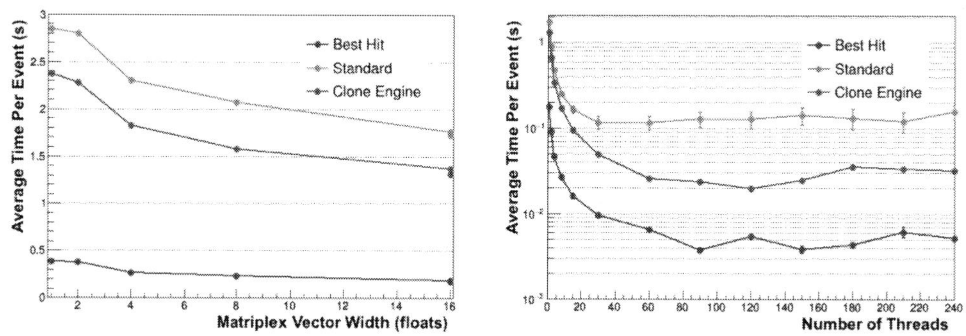

(a) Impact of vector size on performance. Only one thread is enabled.

(b) Impact of the number of threads on performance. The full vector size (16 floating points) is enabled.

Figure 3: Average time per event of 10,000 tracks on Xeon Phi (Knights Corner 7120P) coprocessors. Both plots show results for three different track building approaches. The blue curve is the "best hit" tracking approach. The green and red curves use the full combinatorial approach in track building. The red curve moves the copying of tracks outside of the vectorizable operations.

clock cycle. Therefore, to keep a given physical core busy every clock cycle, one has to schedule two threads per core alternating in instruction execution. A form of hyperthreading is also present on Xeon Phi, yielding a total of 244 hardware threads (logical cores). Overall, a factor of about 30× speedup is observed.

 DOI: 10.1051/epjconf/201715000006

4 Porting KF based algorithms to the GPU

The challenges to port the Kalman Filter to the GPU are similar in principle to those on x86 multi-core platforms. The parallelization and vectorization issues we encounter are only exacerbated by the fine grain parallelism required by graphical processors. In particular, the large core count of GPUs and their thread scheduling policies force a large number of threads to be concurrently scheduled. It also implies that coping with branching is critical, as threads of a *warp* (a group of 32 consecutive threads) execute the same instruction at the same time.

We have followed the same incremental strategy to move onto GPU architectures as the one we adopted for Xeon and Xeon Phi architectures. We first started by studying and choosing a data structure adapted to operations on many small matrices. We then ported our algorithms, starting from track fitting: its simpler nature made it easier to understand the requirements of porting KF based algorithms to GPUs. Armed with this knowledge, we made a first stab at porting track building algorithms. The best hit approach —where only the candidate yielding the best χ^2 is considered at each layer, for each seed —is the natural one to develop, as it introduces a limited number of new problems. The main issue is to understand how to deal with numerous memory indirections. Pursuing again the same stepwise improvements as those done for x86 processors, we next looked at the combinatorial algorithm, where multiple candidates per seed are considered at each layer. This introduces the additional problem of branching.

4.1 Memory Layout

Matriplex has been found to be the most efficient memory layout for GPU, as it naturally offers coalescent accesses. Coalescent accesses occur when, in the case of 4-byte elements, 32 consecutive threads request 32 consecutive elements from global memory. In conjunction with aligned memory accesses, it has the desirable effect of allowing for a single memory transaction instead of 32 separate ones. As a global memory (the large off-chip memory) request takes between 200 and 400 clock cycles to complete, this leads to significant gains. In general, poor performance can be expected unless memory accesses are coalescent. A key difference with CPU Matriplex is the number of matrices grouped within such a data structure. Instead of being tailored to the dimension of a hardware vector unit, it contains as many matrices as possible in order to increase the parallelism exposed to the device. Figure 4 shows some results of a study we conducted to determine the best memory layout, considering that a large part of the KF operations involves multiplying small matrices. In all cases, shared memory is used to reduce the number of global memory transactions, and this has a dramatic impact on the performance, as much as two orders of magnitude. By using the Matriplex data structure, we are able to outperform the finely tuned and batched cuBLAS approach. As a results, all our subsequent work has been based on the GPU version of the Matriplex data structure.

4.2 Track Fitting on the GPU

As mentioned, track fitting is suitable for developing a better understanding of the requirements to obtain performance on an unexplored hardware architecture, such as GPUs. Porting from C++ to CUDA is made tedious by the lack of a library matching the C++ STL and usable from device code. The existing options (for example NVIDIA Thrust [15]) are higher level abstractions to be used from the host and have to be casted to raw pointers when passed to device kernels. The process of optimizing the code so that it runs fast enough is often orders of magnitude harder than the process of rewriting to CUDA. In what follows, we choose the most straightforward way to map computations to the GPU by using one GPU thread per track during fitting.

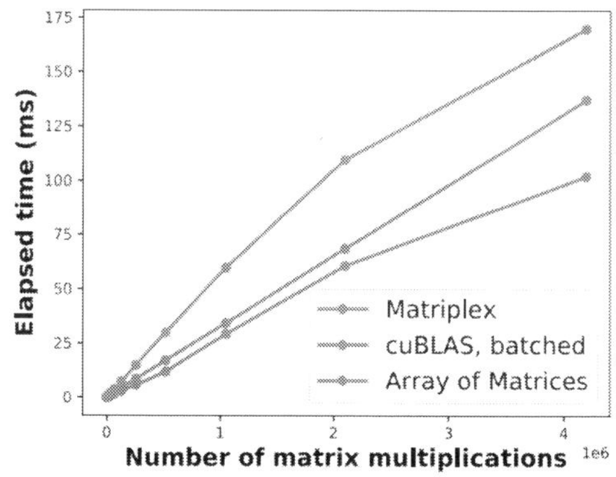

Figure 4: Performance comparison of different memory organizations for multiplying 6×6 matrices. It shows the time as a function of the number of 6×6 matrix multiplications. The orange and purple lines use an array of matrices, while the green line uses the MATRIPLEX data structure.

There is a number of places where optimization can occur. The most common places are the kernels themselves (i.e. the CUDA routines) and the data transfers between the host (CPU) and the device (GPU). We have observed that optimizing kernels brings significant gains. The most beneficial kernel-level optimizations were using read-only caches, merging kernels to limit the launch overhead of many small kernels, and favoring registers over shared memory when possible. These kernel optimizations resulted in a speedup of roughly 8 with respect to a naive implementation. Reorganizing data, in particular to fill MATRIPLEX structures, is costly on the GPU because of the large number of memory indirections it requires. For track fitting, we have chosen to let this reorganization happen on the CPU, which is more suited to this task. By launching kernels asynchronously inside streams, we are able to overlap these reorganizations with actual computations. An additional lesson we learned is the need to fill the GPU by fitting tracks of more than one event concurrently. Indeed, one event is too small to completely fill the GPU by itself. By using multiple CPU threads, each of them associated with a list of events to process, significant improvements are obtained. Figure 5 shows the impact of concurrent fitting. For small MATRIPLEX sizes, the kernels perform poorly due to a low compute intensity. For larger sizes, multiple CPU threads are better able to fill the GPU, and performance improves by a factor of roughly 3.

4.3 Track Building on the GPU

Track building is a more complex problem to deal with on multi- and many-core architectures. We approach it with two different algorithms. The first one, "best hit" considers a single candidate hit for each seed at each layer. The second one is a combinatorial algorithm, keeping track of multiple candidates for each seed at each layer, where candidates are ranked by the number of hits found and by their χ^2. Note that only the "clone engine" approach has been attempted on GPUs: the other combinatorial algorithm involves too many memory copies to be even considered.

Compared to track fitting, the "best hit" approach increases the complexity by adding many memory indirections. For each track, hits within η bins have to be considered. This implies frequent reorganizations into MATRIPLEX data structures, making the overlapping of reorganizations and computations not possible anymore. Reorganizing hits into MATRIPLEX is consequently done on the GPU.

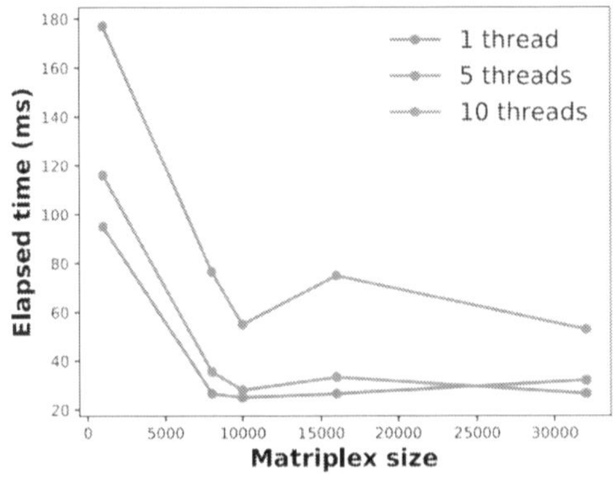

Figure 5: Concurrent fitting processes on the GPU. Different plots show different number of CPU threads concurrently streaming events to the GPU. Time is plotted as a function of the MATRIPLEX size.

As in track fitting, the parallelization relies on having one GPU thread per candidate. Modular and nested data structures can fragment data transfers, even if they are asynchronous, and will in most cases have to be redesigned to a shallower hierarchy. A data structure with multiple layers also forces the inconvenience of having to use additional arrays to keep track of pointers to GPU memory. As a side note, newer versions of CUDA offer unified memory as a way to transparently access data that might be located either in the main RAM or in the GPU global memory. The cost of such a convenience has been greatly reduced over the past years, but the frequency of these transfers dictates a more controlled approach, using the knowledge we have to pre-transfer data asynchronously.

The combinatorial approach generates even more complications when aiming at performant GPU computations. Having multiple candidates per seed at each layer means that conditional branches are encountered frequently. Since warps always schedule the same instruction at a given time, branches serialize the warp execution when the condition differs for each thread of the warp. We address this issue by defining a data structure to store information about multiple candidates. While on x86 processors standard C++ vectors are used, on GPUs the underlying choices are more limited and boil down to C-arrays. Our implementation takes the form of a heap based approach, often used to implement priority queues. Details are shown in Figure 6. A heap is a complete binary tree and can be implemented as a 0-based array where an element at index i has children at indexes $2i + 1$ and $2i + 2$. This data structure uses shared memory to avoid extraneous requests to global memory. At the beginning of each layer, a heap is assigned to each track candidate to store the best suited hits. Each heap is accessed by a single thread, and elements are consecutive in the vertical direction to avoid bank conflicts. At the end of each layer, new overall candidates for the best track need to be found by sifting all the heaps. Figure 7 explains this sifting process. A number of heaps are sorted and their best elements are push-popped into non-sorted heaps. The process is repeated until a single heap remains. The last heap contains information about the bests of all the candidates and is sorted in order to proceed to the next layer. Initially, heaps are filled with guards, allowing a different number of potential hits to be considerate per initial track candidate. The number of potential hits is limited by the statically defined heap size.

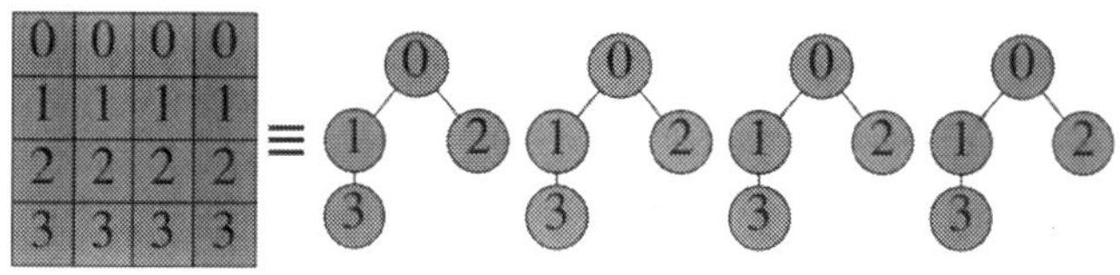

Figure 6: The diagram on the left represents the shared memory layout for storing a seed's candidates. Assuming a maximum of 4 candidates per seed, each column represents a heap to store potential new hits for a track candidate. Each different color represents a different initial track candidate of the seed at a given layer. Numbers represent indexes of the heap's elements. Memory is consecutive in the vertical direction. The diagram on the right shows an equivalent representation using binary trees.

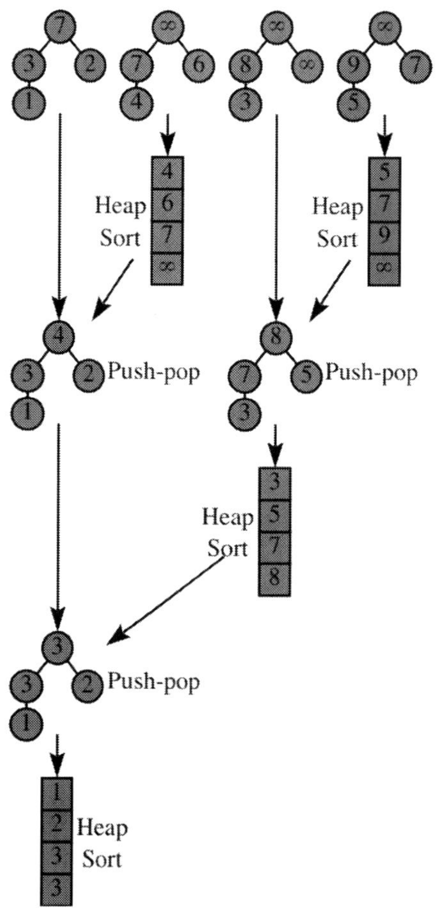

Figure 7: Reducing candidates in shared memory for a single seed. At the end of each layer, the new candidates computed from the starting 4 candidates are sifted to a new set of 4 candidates. ∞ signs represent guards for the heap based algorithm. Numbers are χ^2 values chosen to be illustrative.

A large part of the development effort to port track building to GPUs has been to bring the performance to a level comparable to the one observed on the extensively tuned Xeon version. Results for the "best hit" approach to track building are shown on Figure 8a. Considering the computational time as well as the data transfer, the GPU version is about 3 times faster than a single-threaded execution on a Sandy Bridge (SNB) processor but about 4 times slower than an execution using 24 hyper-threads. The performance comparison is even more severe for the combinatorial version, as shown in Figure 8b. It is a consequence of the large number of synchronizations and branch predictions.

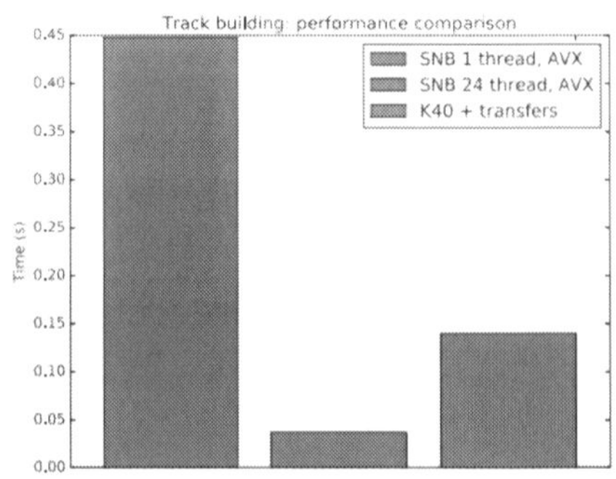

(a) Performance comparison for the "best hit" approach.

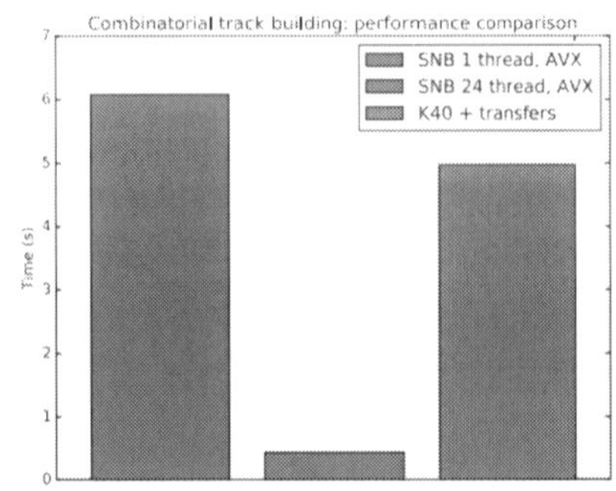

(b) Performance comparison for the combinatoric "clone engine" approach.

Figure 8: Track building performance comparison between the GPU version and the optimized Xeon version. The GPU is a NVIDIA K40 and the Xeon a Sandy Bridge E5-2620

Several future developments can improve the performance of the GPU code. First, concurrent events should be adapted to track building because this will help to fill the GPU's stream multiprocessors more efficiently, as we have seen for track fitting. Next, alternative parallelization patterns need to be studied to completely understand what are the most important performance-driving factors. For instance, using one GPU thread per seed may reduce the amount of synchronization required between threads, but will make concurrent event processing even more critical. Finally, data transfers need to be drastically improved as they now account for almost half of the track building wall time.

5 Conclusion and Outlook

We have made significant progress in parallelized and vectorized Kalman Filter-based tracking on multi- and many-core processors. On Xeon and Xeon Phi, improvements have been made to include a more realistic geometry. These changes were not discussed here as they are not reflected on the GPU side, yet. On all platforms many challenges remain to fully exploit the computational power of these highly parallel architectures. For instance, on the GPU, the bottlenecks we will have to address next are data transfers between the host and the device, as well as branching and synchronization within kernels. The project has produced promising results; however, much work remains.

Acknowledgment

This work is supported by the U.S. National Science Foundation, under the grants PHY-1520969, PHY-1521042, PHY-1520942 and PHY-1120138.

References

[1] CMS Collaboration, Physics Letters **B716**, 30 (2012)
[2] ATLAS Collaboration, Physics Letters **B716**, 1 (2012)
[3] G. Cerati, Proceedings of Science **VERTEX2014**, 037 (2015)
[4] R. Fruehwirth, Nuclear Instrumentation Methods A **262**, 444 (1987)
[5] P. Billoir, Computer Physics Communications **225**, 352 (1984)
[6] R. Mankel, Nuclear Instruments and Methods in Physics Research A **395**, 169 (1997)
[7] V. Haylo, P. LeGresley, P. Lujan, V. Karpusenko, and A. Vladimirov, Journal of Instrumentation **9**, P04005 (2014)
[8] G. Cerati *et al.*, Journal of Physics: Conference Series **608**, 012057 (2014)
[9] G. Cerati *et al.*, Journal of Physics: Conference Series **664**, 072008 (2015)
[10] G. Cerati *et al.*, "Kalman-Filter-Based Particle Tracking on Parallel Architectures at Hadron Colliders" (2015), NSS-MIC
[11] G. Cerati *et al.*, Journal of Physics: Conference Series (2016), https://arxiv.org/abs/1702.06359
[12] G. Cerati *et al.*, EPJ Web Conf. **127**, 00010 (2016)
[13] *Intel VTune Amplifier*, `https://software.intel.com/en-us/intel-vtune-amplifier-xe`
[14] *Intel Thread Building Blocks*, `https://www.threadingbuildingblocks.org/`
[15] *NVIDIA Thrust*, `https://developer.nvidia.com/thrust`

The track finding algorithm of the Belle II vertex detectors

Tadeas **Bilka**[1], Nils **Braun**[2], Giulia **Casarosa**[3,4,5], Oliver **Frost**[6], Rudolf **Frühwirth**[7], Thomas **Hauth**[2], Martin **Heck**[2], Jakub **Kandra**[1], Peter **Kodys**[1] Peter **Kvasnicka**[1], Jakob **Lettenbichler**[7], Thomas **Lück**[3,4,8,a], Thomas **Madlener**[7], Felix **Metzner**[2], Moritz **Nadler**[7], Benjamin **Oberhof**[3,4], Eugenio **Paoloni**[3,4], Markus **Prim**[2], Martin **Ritter**[9], Tobias **Schlüter**, Michael **Schnell**[10], Bjoern **Spruck**[5], Viktor **Trusov**[6], Jonas **Wagner**[2], Christian **Wessel**[10], and Michael **Ziegler**[2]

[1]*Charles University, Prague (Czech Republic)*
[2]*IEKP, Karlsruhe Institute of Technology, Karlsruhe (Germany)*
[3]*I.N.F.N. sezione di Pisa, Pisa (Italy)*
[4]*Università di Pisa, Pisa (Italy)*
[5]*Institut für Kernphysik Johannes Gutenberg-Universität, Mainz (Germany)*
[6]*Deutsches Elektronen-Synchrotron, Hamburg, (Germany)*
[7]*Institute of High Energy Physics, Vienna (Austria)*
[8]*Scuola Normale Superiore di Pisa, Pisa (Italy)*
[9]*Excellencecluster Universe, Ludwigs-Maximillians-University (Germany)*
[10]*Physikalisches Institut Rheinische Friedrich-Wilhelms-Universität, Bonn (Germany)*
[11]*Excellencecluster Universe, Ludwigs-Maximillians-University, Munich (Germany)*

Abstract. The Belle II experiment is a high energy multi purpose particle detector operated at the asymmetric e^+e^- - collider SuperKEKB in Tsukuba (Japan). In this work we describe the algorithm performing the pattern recognition for inner tracking detector which consists of two layers of pixel detectors and four layers of double sided silicon strip detectors arranged around the interaction region. The track finding algorithm will be used both during the High Level Trigger on-line track reconstruction and during the off-line full reconstruction. It must provide good efficiency down to momenta as low as 50 MeV/c where material effects are sizeable even in an extremely thin detector as the VXD. In addition it has to be able to cope with the high occupancy of the Belle II detectors due to the background. The underlying concept of the track finding algorithm, as well as details of the implementation are outlined. The algorithm is proven to run with good performance on simulated $\Upsilon(4S) \to B\bar{B}$ events with an efficiency for reconstructing tracks of above 90% over a wide range of momentum.

1 Introduction

Belle II is a multipurpose detector which will be operated at the asymmetric B-Factory SuperKEKB (Japan). The unprecedented instantaneous luminosity of up to $8 \times 10^{35}\,\mathrm{cm}^{-2}\,\mathrm{s}^{-1}$ provided by the accelerator together with the soft spectrum of the charged particles to be traced (typical momentum

[a]e-mail: lueck@pi.infn.it

350 MeV/c and with the high occupancy (1.5%) foreseen in the inner silicon layers will challenge the sub-detectors of Belle II and the track finding algorithms.

Track position information close to the interaction point is provided by the vertex detector (VXD) which consists of 2 layers of pixel detectors (PXD) and 4 layers of double sided silicon strip vertex detectors (SVD) whose overall thickness and dead material is kept at a minimum to reduce the effects of multiple Coulomb scattering.

The track finding code for the VXD of Belle II implements in an efficient way the Sector Map concept originally proposed by Rudolf Frühwirth and originally implemented by Jakob Lettenbichler [1].

The typical event recorded by the VXD will be dominated by random hits produced by beam background (order of 500 pixels hit per event per sensor on the inner layer of the PXD, 20 GByte/s required to readout the whole PXD). In this harsh environment the pattern recognition algorithm for the VXD has to be capable to efficiently and quickly recognize the 11 long lived charged particles emerging as decay products of a typical $\Upsilon(4S)$ event.

The track finding algorithm presented here will be used both during the final reconstruction of the event and during the fast reconstruction occurring on the High Level Trigger (HLT) for the definition of the Regions Of Interest on the PXD sensors used to reduce the PXD data stream to a manageable level [2]. This latter task put tight constraints on the reliability and time consumption of the track finding algorithm since an eventual crash or malfunction of the HLT process will immediately translate in a permanent loss of data and since the time left for the silicon vertex identification is in the ball park of a few milli-seconds per event.

This paper presents the main concepts of the algorithm, some details of its implementation together with its current performances.

2 The Belle II detector and the inner tracking devices

The Belle II detector is described in detail elsewhere [3]. Here are described only the inner silicon tracking devices for which the track finding algorithm is designed. The inner tracking detector consists of two layers of DEPFET pixel detectors (PXD) [4], which are arranged around the beam at the radii of 14 mm and 22 mm from the interaction point. Its inner layer consists of 8 sensors with dimension $15 \times 90\,\text{mm}^2$ while its outer layer consist of 12 modules of the size of $15 \times 123\,\text{mm}^2$. The silicon has been thinned down to $75\,\mu\text{m}$ to reduce the radiation length. The angular acceptance starts at a polar angle of $17°$ in the forward direction and ends to $150°$. The size of the pixels is $50 \times 50\,\mu\text{m}^2$ for the inner layer and $50 \times 75\,\mu\text{m}^2$ for the outer layer, respectively.

The PXD is surrounded by the Silicon Vertex Detector (SVD) [5] which consists of 4 layers of double sided silicon strip detectors positioned at radii of 38 mm, 80 mm, 115 mm, and 140 mm from the interaction point. The sensors have readout pitches in the $r - \phi$-direction (p-Side) between 50 μm and 75 μm, and between 160 μm and 240 μm in the z-direction (n-Side). To increase the resolution of the reconstructed position a floating strip is present in-between two readout strips. The sensors are arranged in ladders which are made of 2, 3, 4, and 5 sensors per ladder for layer 3, 4, 5, 6 (counting starts from the first PXD layer), respectively. The number of ladders per layer is 7, 10, 12, 16 going from the inner to the outer layer of the SVD, respectively. To reduce the amount of material in the forward region the shape of the ladders in layers 4, 5, and 6 are such to form a peculiar "lamp shade" geometry (cfr. Fig. 1) in which the forward sensors for layers 4, 5, and 6 are trapezoids slanted with respect to the beam pipe.

Both SVD and PXD are operated within a superconducting solenoid which provides a magnetic field of 1.5T field strength.

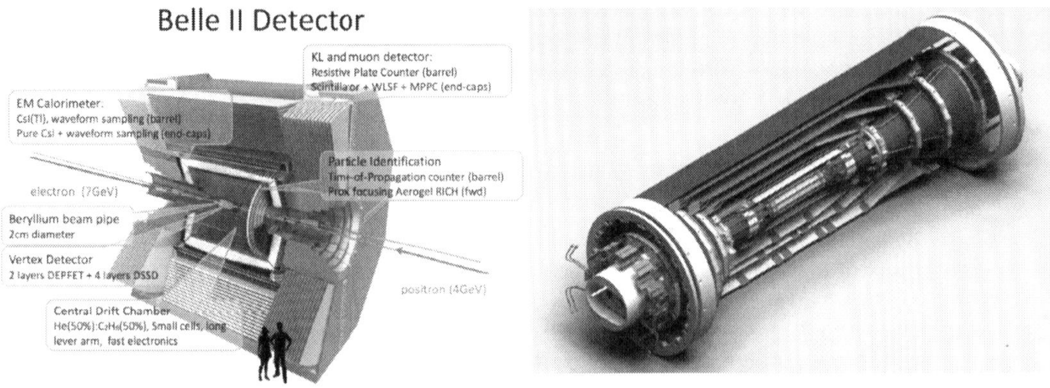

(a) The Belle II detector.

(b) Schematic of the Vertex detectors.

Figure 1: Shown is a schematic of the Belle II detector (a), and a schematic of the vertex detectors of Belle II (b).

3 Track finding concept

Due to the nature of the B-meson decays the track finding has to deal with very low momentum particles. The most probable momentum of the long lived decay products of the B is around 350 MeV/c moreover the track finding algorithm must efficiently work down to momenta of around 50 MeV/c for which the only tracking informations are coming from the VXD. An additional challenge is the extremely high background from by the pairs productions process $e^+e^- \rightarrow e^+e^-e^+e^-$ whose large cross section will rise the occupancy in the PXD inner layer to 1.5 %. The concept of the track finding for the Belle II experiment was chosen to efficiently cope with these requirements in this unprecedently harsh environment: track seeds formed using SVD and DCH hits are used to define Regions Of Interest (ROI) on the PXD. Just the PXD hits inside these ROIs together with additional PXD clusters selected by a Neural Network according to their position, charge, multiplicity and shape will be written on disk for the final event reconstruction.

Additional ROIs will be provided by the DATCON (an FPGA based tracking device that will use only SVD hits and that will run in parallel with respect to the HLT) that is described in details in another proceeding of this very same conference.

The finding of track candidates works in several steps which will be sketched in the following sections. These steps are, in this very order, the combination of two hits to produce segments, the combination of segments into track candidates, the quality estimation of the track candidates and finally the removal of random combinations of hits (that is, fake tracks) from the list of track candidates to be used for the final Kalman fit.

The implementation of the track finding algorithm for the Belle II vertex detectors is implemented in C++ 11 and is part of the Belle II analysis software framework [6].

3.1 The sector map concept.

The Sector Map concept was introduced by Rudolf Früwirth and implemented by Jakob Lettenbichler [1] within the Belle II collaboration. Its unique feature is to learn from a training sample of Monte

Carlo (MC) tracks the geometry of the detector. This is particularly appealing in Belle II since the VXD lacks trivial geometrical discrete symmetry.

The training sample should represent the momentum spectrum and particle type composition (due to differences in interactions with the material for different particle types) expected for the actual data sample. For the results presented in this document a training sample of simulated $\Upsilon(4S) \to B\bar{B}$ events have been used.

At the heart of the concept of sector map is the partition of the sensors surface in Sectors S_i. Each sector S_i is connected in a so called "friendship relation $\mathcal{F}_{ij}$" to a sector S_j if trajectories are passing through S_i and S_j without crossing any other sector in between. If we limit our analysis to the outgoing arm of the track coming from the interaction point, it turns out that the set $\{\mathcal{F}_{ij}\}$ defines a cycle free graph on the set of nodes $\{S_i\}$ (that is, $\{S_i\}$ is the set of nodes and $\{\mathcal{F}_{ij}\}$ is the set of edges). Moreover the curvilinear coordinate on the trajectories defines a natural direction of each edge of this graph, so that in the end the trajectories induce on $\{S_i\}$ a directed sortable graph structure that will be exploited in the following. A subset of this graph is represented in Fig. 2.

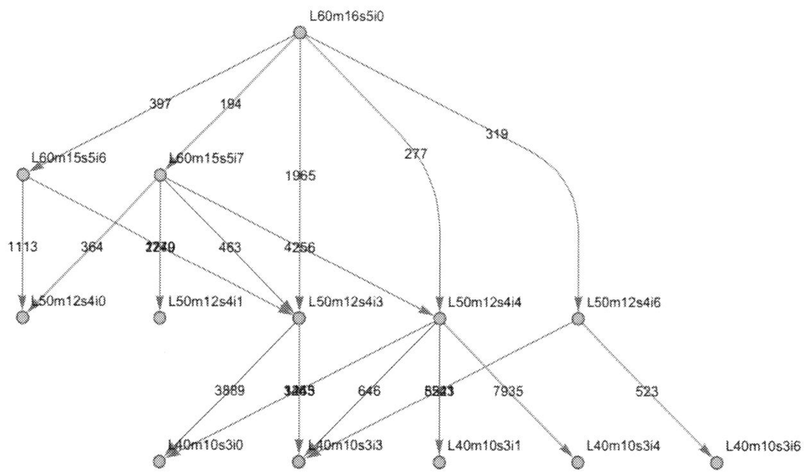

Figure 2: A subset of the graph defined by friendship relations among sectors. The format of the sector name is L *ll* m *xx* s *yy* i *zz* where *ll* is the layer and sub layer id, *xx* is the module number, *yy* is the sensor number and *zz* is the sector number. The number on the directed edge corresponds to the number of tracks in the training samples connecting the two sectors.

Sector S_i is defined as outer with respect to S_j if the trajectories connecting them are going from S_j to S_i. With these natural definition the outer sectors are on the geometrical outer layers of the tracking device. A sample of MC tracks are used for defining the friend relations among sectors. The MC events used for the definition of the friend relations are produced with a full Geant 4 detector simulation. In the current version of the code both the number and the position of the corners of the sectors can be set individually for each sensor, separately for u and v-direction. This allows to chose a different setting for the sectors in different detector regions for example to cope with different level of backgrounds found in these regions. The results presented in the following are obtained partitioning the sensors in 9 sectors arranged in a 3×3 array.

The first step of the reconstruction is to assign each detector hit to the sector to which he belongs (this operation is linear in the detector occupancy). At present SVD clusters on the two orthogonal views of a sensor are preliminary combined in a single Space Point (SP) which represents the 3D position in the global coordinates system. The SPs are stored in a container whose access key is the Sector S_i on which the SP belongs. A sector containing one or more SP is defined as an "active sector".

The next step is to find compatible combinations of two hits (in the following called segments). A loop over the active sectors is performed and for each hit belonging to a sector S_i the second hit is searched only in active sectors S_j which are friends of the sector S_i. The approach of considering only pairs of friends sectors drastically reduces the number of possible two-hit combinations moreover, since the average occupancy of the sector is order one SP / sector under pessimistic background assumptions just in the inner SVD layer, the computational complexity of this step is still linear with the number of SP (at least up to reasonable background running conditions).

For each friendship relation $\mathcal{F}_{ij}$ a set of geometrical requirements is defined to be fulfilled by the segment s to be accepted and used in the next steps of the reconstruction. The geometrical requirements are expressed as a set of inequalities on several geometrical variables $v_\alpha(s)$. Due to the complicated geometry of the detector it is not sufficient to set the limits for the $v_\alpha(s)$ globally. For that reason the upper and lower bound for each $v_\alpha(s)$ are set individually for each friendship relation. The inequalities used so far have the simple form $\min_{ij\alpha} \leq v_\alpha(s) \leq \max_{ij\alpha}$ where the min and max values allowed for $v_\alpha(s)$ do depend on the indexes of the two friends sectors i and j and their value is tuned using the training sample used also for the definition of the sector connection. The currently implemented geometrical variables for the characterization of the segment are:

- the squared 2D distance between the two SPs in the x-y-plane (perpendicular to the B-field),
- the squared distance between the two SPs in 3D,
- the distance between the two SPs in the direction of the B-field,
- the angle between the direction of the B-field and the direction defined by the two SPs,
- and the normalized distance in 3D between the two SPs.

The segments fulfilling these requirements are accepted and stored in a container whose access key is the friendship relation $\mathcal{F}_{ij}$ where S_i (S_j) is the sector containing the outer (inner) SP of the segment. A friendship relation is defined active if there is at least one segment associated to it. The accepted segments are passed to the next step in which they are combined into SP triplets t. For this step the graph structure induced by trajectories on the set $\{\mathcal{F}_{ij}\}$ is again exploited. A graph is built in which the nodes are the set of friendship relation $\{\mathcal{F}_{ij}\}$ and the edges are represented by the so called "neighbouring relation" $\{\mathcal{N}_{ijk}\}$. The edge $\mathcal{N}_{ijk}$ connects $\mathcal{F}_{ij}$ to $\mathcal{F}_{jk}$. Two friendship relations $\mathcal{F}_{ij}$ and $\mathcal{F}_{jk}$ are defined neighbouring if there are trajectories in the MC sample that are defining at the same time the two friendship relations $\mathcal{F}_{ij}$ and $\mathcal{F}_{jk}$.

To produce the triplets the algorithm loops over the set of active friendship relations $\{\mathcal{F}_{ij}\}$ and for each segment s_1 associated to $\mathcal{F}_{ij}$ the algorithm looks for segments $\{s_2\}$ associated to the neighbouring friendship relation $\mathcal{F}_{jk}$. Two segments s_1 and s_2 are combined in a triplet t if the hit on the inner sector of s_1 is the very same of the one on the outer sector of s_2 and if they fulfill a set of geometrical requirements.

On the same footing of the segment selection a few geometrical variables to characterize the triplet are defined and it is required that these variables belong to intervals that are a function of the neighbouring relation $\mathcal{N}_{ijk}$. The size of the intervals, also in this case, is automatically tuned on the training MC sample. The following filters are currently implemented:

- the angle in 3D between the two segments,
- the analogue angle defined the x-y-plane (plane perpendicular to the B-field),

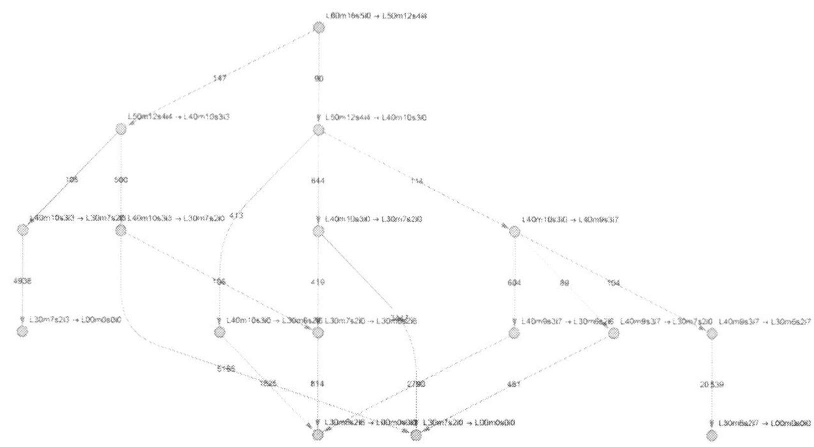

Figure 3: A subset of the graph defined by neighbors relations among friendship relations. The format of the sector name is the same used in Fig. 2.

- the angle in the R-z-direction,
- the distance of the center of the circle in the x-y-plane, defined by the 3 hits, w.r.t the interaction point,
- the change of the slope between the first and the second segment in the R-z-plane,
- the change in slope in the R-z-plane normalized over the distance between the respective hits,
- the change in the distance between the first and second hit normalized over the distance in z with respect to the second and third hit,
- the transverse momentum calculated from the circle defined by the three hits in the x-y-plane,
- and finally the radius of the circle the three hits define in the x-y-plane.

The framework also has the possibility to apply filters on four hit combinations. At the moment this option is not used as the filtering of track candidates containing at least four hits can be done more efficiently by the implemented quality estimators (see Section 3.3 for more details).

The friendship relations together with the neighbouring relations and the upper and lower limits for all the filters are stored in a lookup-table called SectorMap which memory footprint is order of a few tens of MBytes.

3.2 The cellular automaton

The next task the track finding algorithm has to perform is to connect the segments that form the triplets into track candidates. For this task a cellular automaton algorithm is used [7]. In this context a cell corresponds to a segment. Two cells are neighbors if and only if the two corresponding segments s_1 and s_2 are linked in a triplet. The sorting of the friendship relations induces a sorting on the cells. A cell s_1 is defined as"outer" with respect to its neighbors s_2 if s_1 belongs to a friendship relation that is outer with respect to the friendship to which belongs s_2. The state of the cell is defined by an integer number called the *depth* and a boolean flag called *toBeUpdated*.

At the beginning the *depth* of each cell is set to 0 and the flag *toBeUpdated* is set to false. The cellular automaton evolution is governed by these two steps:

- Check step: all the cells that have all theirs inner neighbors in a state with the same *depth* are marked by setting to true the flag *toBeUpdated*
- Update step: for all the cells whose flag *toBeUpdated* is true we increase by one the *depth* and we reset to false the *toBeUpdated* flag.

The two steps are repeated until the state of all cells reaches a steady state. After the cellular automaton has converged the track candidates are generated by starting with the cells having the highest depth and collecting recursively all inner neighbors. If a cell has two inner neighbors the track candidate is duplicated and both branches are retained. Figure 4 shows an illustration of a reconstructed event after the the cellular automaton has converged.

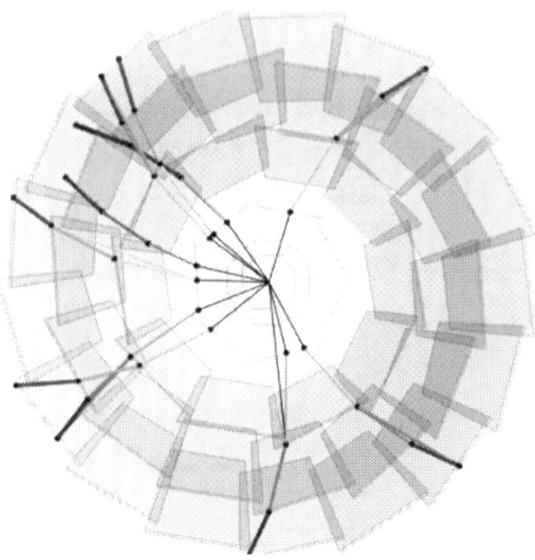

Figure 4: Shown is a toy event after the reconstruction using the cellular automaton. The different colors represent the different states the cells (segments) are in. (plot taken from Ref. [1]).

3.3 Quality estimators

To perform the cleanup of the track candidates a quality indicator is assigned to each track candidate. The estimation of this quality indicator has to be very fast to keep the overall computing resource consumption small as it has to be performed on many track candidates. The range of this quality estimator is by definition between 0 and 1 where 1 is the best quality and the 0 is the worst.

Currently there are three different methods implemented to estimate the quality indicator: a simple circle fit, a Tiplet Fit and a helix Riemann fit [8]. The P-value of the Chi-square is used as quality indicator.

3.4 Track selection

After the previous steps there exist still a large fraction track candidates which consist of random combinations of hits in the detector. The final step in the track finding procedure is to find among all the track candidates, found in the previous steps, a clean subset of track candidates. In this context clean means that no two track candidates share a common hit with each other.

Two methods for the cleanup are currently implemented. One uses a Hopfield Network [9]. For this purpose a Neuronal Network is constructed where the nodes are the track candidates. The weights of the connections between the nodes are determined by the quality indicator (see Section 3.3) and the sharing (or not sharing) of common hits between the two track candidates.

The second algorithm implemented is the so called Greedy algorithm. This algorithm removes all track candidates which share a hit with the track candidate with the highest quality estimator. After these track candidates are removed the procedure is repeated, on the remaining track candidates, with the track candidate with next highest quality estimator until no track candidates are removed any more, which is equivalent with reaching the track candidate with lowest quality estimator.

4 Performances and conclusions

The performance of the track finding algorithm is evaluated using fully Geant 4 simulated $\Upsilon(4S)$ Monte Carlo events (MC). For this purpose a dedicated tool, the Ideal Track Finder (ITF), combines all reconstructed hits which belong to true charged particle into a track candidate by exploiting the MC true informations.

The track finding efficiency for the algorithm at hand is then evaluated as the fraction of the number of "good" tracks found by the pattern recognition algorithm described so far over the number of track candidates found by the ITF. We require for a track to be defined as "good" that at least 66% of their clusters are produced by the same MC particle.

The results are summarized in Fig 5, the efficiency above $p_t \sim 50$ MeV/c exceeds 80%. The fake rate, before the final Kalman filter selection, is foreseen to be 0.06%. Figure 6 shows the fake rate as a function of the azimuth angle ϕ.

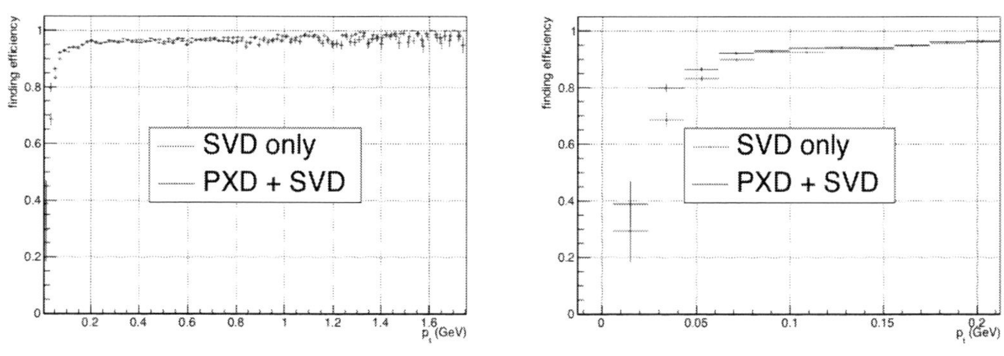

Figure 5: Track finding efficiency relative to an ideal track finder as a function of the transverse momentum using only the hits detected by the SVD and using all the informations coming from the VXD (PXD + SVD). On the left-hand side the results for the efficiency are shown in the range from p_t = 0 GeV/c to p_t = 1.75 GeV/c. On the right-hand side the same results are shown in the range from p_t = 0 GeV/c to p_t = 200 MeV/c.

The performance results shown here have been evaluated on 10^4 simulated events of the process $\Upsilon(4S) \rightarrow B\bar{B}$ where the B-mesons are decaying into a generic final state. For the results shown a setting of 3 by 3 sectors per sensor has been chosen. The quality estimation of the track candidates is performed by the Circle fit (see Section 3.3) and the selection of the final track candidates is performed by the Greedy algorithm (see Section 3.4).

DOI: 10.1051/epjconf/201715000007

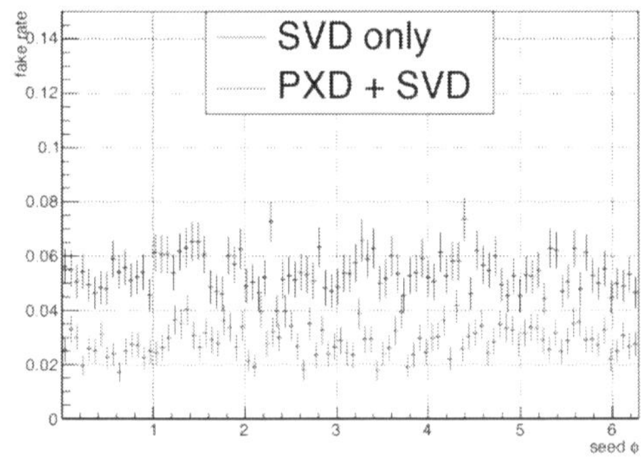

Figure 6: Fraction of fake tracks (random combination of hits) found by the track finding algorithm as a function of the azimuth angle ϕ using only the hits detected by the SVD and using all the informations coming from the VXD (PXD + SVD).

The code was extensively tested on a real working system during a test beam in DESY in which a slice of the whole VXD system (i.e. 2 PXD and 4 SVD ladders) was read out by a scaled version the Belle II DAQ and HLT system. The code was able to sustain a trigger rate of a few kHz and to produce ROI for the PXD.

The start of the Beast II data taking is foreseen in late February 2018, time by which the parameters of the algorithms will be fully tuned.

References

[1] J. Lettenbichler, Ph.D. thesis, Technische Universität Wien (2016)
[2] T. Bilka et al., IEEE Trans. Nucl. Sci. **62**, 1155 (2015), 1406.4955
[3] T. Abe et al. (Belle-II) (2010), 1011.0352
[4] H.G. Moser (DEPFET), Nucl. Instrum. Meth. **A831**, 85 (2016)
[5] D. Dutta et al., JINST **12**, C02074 (2017)
[6] T. Schlüter (Belle-II Software Group), PoS **Vertex2014**, 039 (2014), 1411.3485
[7] A. Glazov, I. Kisel, E. Konotopskaya, G. Ososkov, Nucl. Instrum. Meth. **A329**, 262 (1993)
[8] A. Strandlie, R. Fruhwirth, Nucl. Instrum. Meth. **A480**, 734 (2002)
[9] R. Fruhwirth, Comput. Phys. Commun. **78**, 23 (1993)

The design and simulated performance of a fast Level 1 track trigger for the ATLAS High Luminosity Upgrade

Mikael **Mårtensson**[1,a], on behalf of the ATLAS Collaboration

[1] *Institute of Physics and Astronomy, Uppsala Universitet*

Abstract. The ATLAS experiment at the High Luminosity LHC will face a fivefold increase in the number of interactions per bunch crossing relative to the ongoing Run 2. This will require a proportional improvement in rejection power at the earliest levels of the detector trigger system, while preserving good signal efficiency. One critical aspect of this improvement will be the implementation of precise track reconstruction, through which sharper trigger turn-on curves can be achieved, and *b*-tagging and tau-tagging techniques can in principle be implemented. The challenge of such a project comes in the development of a fast, custom electronic device integrated in the hardware based first trigger level of the experiment. This article will discuss the requirements, architecture and projected performance of the system in terms of tracking, timing and physics, based on detailed simulations. Studies are carried out using data from the strip subsystem only or both strip and pixel subsystems.

1 Introduction

ATLAS is an experiment at the Large Hadron Collider (LHC) aimed at studying the Standard Model of particle physics and to search for physics beyond the Standard Model by analysing high energy proton-proton interactions. The ATLAS detector consists of a number of subdetectors. Moving from the interaction point outwards they are: a tracker that consists of a silicon pixel detector, a silicon strip detector, and a straw tube detector; the electromagnetic and hadronic sampling calorimeters, which use liquid argon and plastic scintillators as active medium; and finally the muon spectrometer that consists of monitored drift tubes, cathode-strip chambers, resistive plate chambers, and thin gap chambers [1].

The High Luminosity LHC (HL-LHC) is scheduled to start running in 2026. The upgrade is expected to increase the instantaneous proton-proton luminosity to $7.5 \times 10^{34}\ \mathrm{cm}^{-2}\ \mathrm{s}^{-1}$, which translates to a sevenfold increase in the number of interactions per bunch crossing, so-called *pile-up*, compared to the LHC design luminosity. The ATLAS detector will be upgraded as well, in a project named the *Phase-II upgrade*, with installation scheduled to take place between 2024 and 2026. The upgrade will include a completely new tracker, as described in section 2.2, upgrades to the trigger and readout electronics, and replacements of parts of the calorimeter and muon spectrometer.

The vast amount of data generated by the increased luminosity puts strict demands on the ATLAS trigger system. Due to its high energy frontier physics program, ATLAS bases its trigger selection on

[a] e-mail: mikael.martensson@cern.ch

high transverse momentum (p_T) lepton selection organised in two levels: one hardware (first-level) and one software. The left plot in figure 1 shows the ATLAS acceptance fraction of a few processes as a function of muon p_T threshold. From it, it is clear that the current trigger p_T threshold of 20 GeV needs to be maintained in order to take full advantage of the increase in luminosity. At the same time, trigger rates need to be kept at a reasonable level to avoid saturating the readout bandwidth. This is shown in the right plot of figure 1, where we see that the first-level electromagnetic trigger rate for electrons has a strong dependence on p_T [2]. As an example, to limit the rate to 20 kHz one would have to increase the threshold to 40 GeV. The ATLAS Phase-II upgrade will introduce a hardware track trigger that can help to control the trigger rate by exploiting extremely good tracking resolution to increase the signal-to-background ratio.

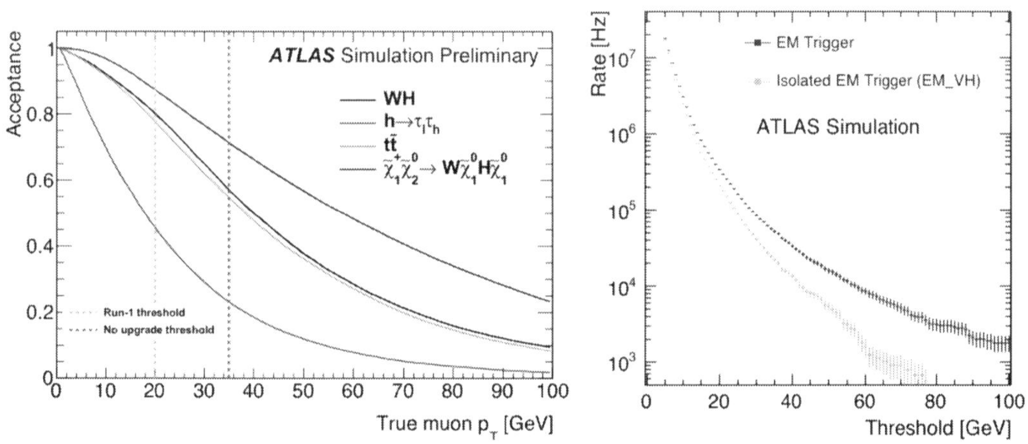

Figure 1: The impact of p_T threshold on acceptance fraction of a few processes (left) and trigger rate for the Level-0 electron trigger (right) for ATLAS at the HL-LHC [2, 3].

2 The ATLAS Phase II Upgrade

The ATLAS detector will be upgraded to make the most out of the HL-LHC. Large parts of the tracker will reach their end of life shortly after 2020 and must be replaced. Therefore, a project to develop a complete replacement of the tracker is in progress. This opens up the possibility of using data from the tracker in the trigger, as described below.

2.1 Trigger Architecture

The ATLAS trigger will consist of a hardware trigger, which uses hardware on and close to the detector, followed by the Event Filter, which is a software trigger running on a computer farm. Current hardware trigger systems, based on calorimeter and muon information, will change to cope with higher trigger rates and will represent the so-called L0 system. The hardware trigger will operate in a single-level (L0) mode but retain the option to evolve to a split-level (L0/L1) trigger. The architecture is presented in figure 2.

The regional hardware track trigger provides tracking in Regions of Interest (RoIs), which are identified by the L0 muon and electromagnetic calorimeter triggers and expected to cover approximately 10 % of the detector volume. In the single-level mode, it assists the Event Filter and is called

EFTrack. In the split-level mode, it contributes to the event rejection before the Event Filter, in the so-called L1 trigger, and is called L1Track. The hardware is the same in both cases, but L1Track has a strict latency requirement and gains speed by increasing parallelism at the expense of reduced momentum acceptance, as explained in section 3.1.1. A comparison of the two working modes is presented in table 1. The Event Filter will also be assisted by FTK++, which provides full-event tracks at a rate of 100 kHz down to a p_T of 1 GeV. FTK++ will not be discussed in this article; see Ref. [4] for further details.

Table 1: Specifications for the regional track trigger when running as an Event Filter co-processor (EFTrack) and as an extra hardware trigger level (L1Track).

Trigger	Latency requirement	L0 rate [MHz]	Trigger threshold [GeV]
EFTrack	$\mathcal{O}(s)$	1	2
L1Track	30 μs	2–4	4

In the single-level mode, the muon and calorimeter triggers can issue a Level-0 Accept (L0A), which initiates full readout of all detectors at 1 MHz. In the split-level mode, L0A is running at a rate of 2–4 MHz and the Inner Tracker (ITk) modules in RoIs specified by the L0-trigger are read out at this rate following a Regional Readout Request (R3). The R3 data is sent to L1Track, which participates in the L1 decision (L1A) to start the full detector readout and Event Filter. The total L1 data rate consists of 10 % of the R3 data plus the Level-1 Accept (L1A) data. If the maximum total readout rate for the ITk is fixed at 1 MHz, then 600–800 kHz is left for L1A, since 200–400 kHz is expected to be consumed by R3, which is prioritised over L1A.

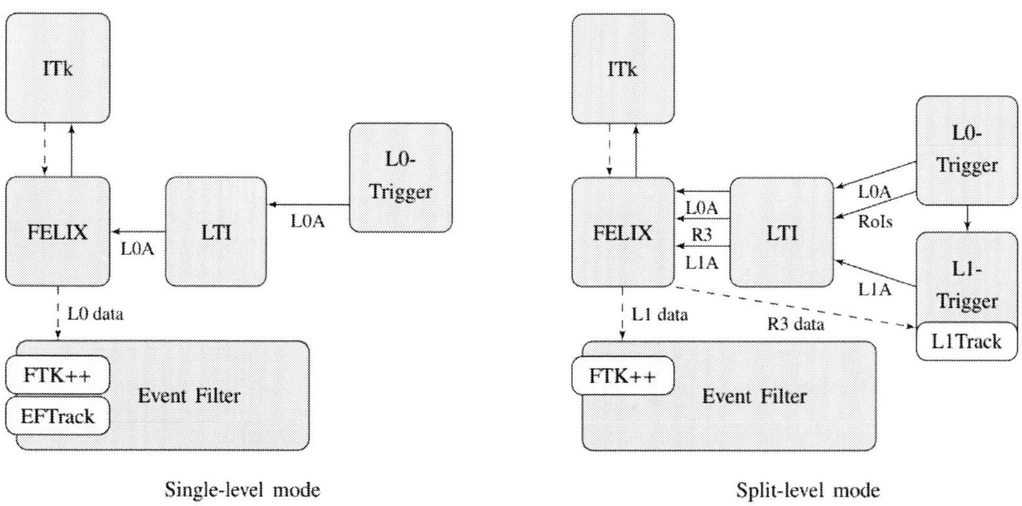

Figure 2: Logical schematic of the two trigger options for the ATLAS Phase-II upgrade. FELIX is the Front-End LInk eXchange and LTI is the Local Trigger Interface. See Ref. [5] for further details.

2.2 The Inner Tracker

The current ATLAS tracker will be completely replaced during the Phase II upgrade. The new ITk will be all-silicon and consist of pixel detectors close to the beam pipe and strip detectors at larger radii. The final layout is not decided yet, but the proposed layouts typically have a barrel with 5 pixel layers and 4 strip stereo-layers (closely-spaced layers, slightly inclined with respect to each other) and an endcap with coverage up to a pseudorapidity (η) of 4.

An important factor for the hardware track trigger is the time it takes to read out data from the ITk. If it takes too long, the buffers in the frontend electronics will get overfilled, which results in loss of data. The left plot in figure 3 shows the result of discrete event simulations of the latency at which 99 % of the event data is read out for the strip endcap in the single-level trigger mode, as a function of L0 rate. The baseline L0 rate is set to 1 MHz, which is below the point on the curve where the latency starts to rapidly increase. At 1 MHz the latency is approximately 6 μs for the module with highest occupancy.

The readout latency in the split-level L0/L1 mode is presented in the right plot of figure 3 for the module with the highest occupancy. The L1A latency is below 8 μs for a L1A rate of 800 kHz with a L0A rate of 2 MHz or a L1A rate of 600 kHz with a L0A rate of 4 MHz. In both cases the R3 latency is below 3 μs, leaving time to process data for a track trigger decision.

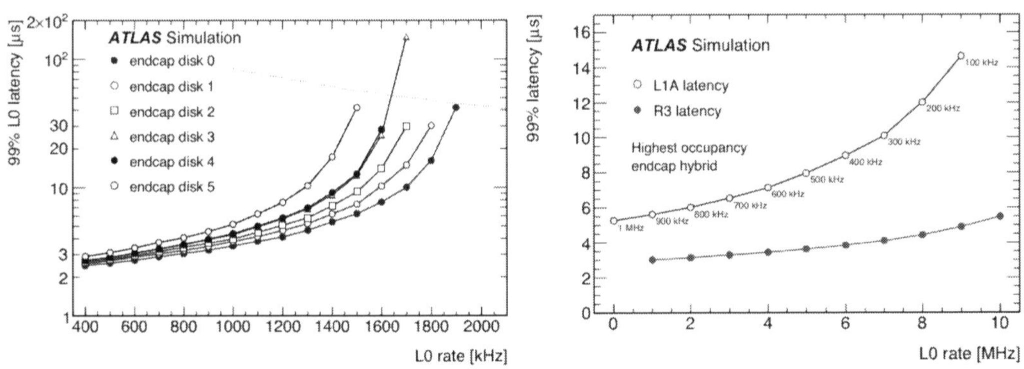

Figure 3: Latencies for the inner tracker strip detector as a function of L0 rate for the so-called *extended* layout. The left plot is showing the latencies for the endcap modules, including the module with highest occupancy, in the single-level L0 trigger mode. The red dotted line indicates the boundary above which frontend buffers are starting to get overfilled. The right plot is showing the highest-occupancy endcap module in the split-level L0/L1 trigger mode. The blue numbers indicate the corresponding L1A rate. The figures are from Ref. [5] where further information can be found.

3 The Regional Track Trigger

The L1Track/EFTrack trigger will provide fast regional hardware track triggering. As described in section 2.1, the regional track triggering will either run as part of an extra trigger level after L0 (L1Track) or as part of the Event Filter (EFTrack). An overview of the system is presented in figure 4. The system conceptually consists of three parts: pattern matching using Associative Memory (AM) chips [6], track fitting of the full-resolution hits using Field Programmable Gate Arrays (FPGAs), and selection of the best track candidates based on track parameters and the χ^2 of the fit. The results presented in this section are for the ITk layout described in the ATLAS Phase-II upgrade Letter of Intent (Ref. [2]).

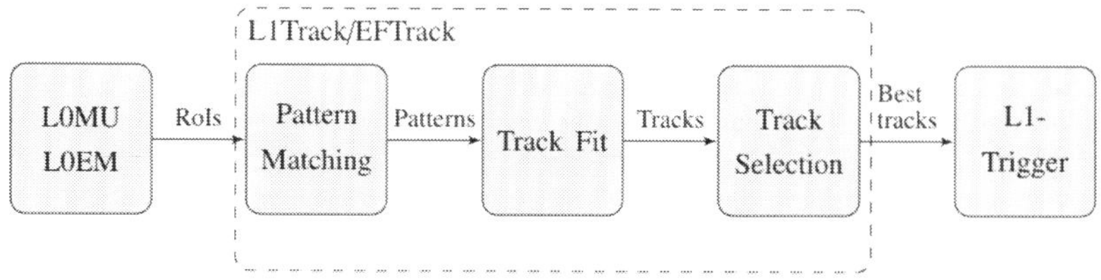

Figure 4: Logical schematic of the L1Track/EFTrack trigger and the objects passed between the system components.

3.1 Pattern Matching

The pattern matching step selects which hits to pass on to the track fitter. This is done using custom made AM ASICs that match the hits read out from the detector to predefined hit patterns from simulated tracks [7]. An alternative method using the Hough transform has also been studied.

3.1.1 Associative Memory

At each clock cycle the AM chip compares the input combination of hits with all the predefined patterns stored in the AM, and outputs the address of the matched patterns, if any. In the regional hardware track trigger, pixels and strips are combined into coarser-resolution *super-strips* and simulated tracks of muons are used to generate patterns of super-strip hits. Data from the detector are sent to the AM chip and if it matches a pattern the associated ITk hits are sent to the track fitter.

The trigger uses 8 ITk layers, which can be a combination of strip and pixel layers. However, the occupancy of the innermost pixel layers is too high to be of use in the track trigger. It is assumed that each AM chip can contain half a million patterns, so that one million patterns can be stored for each $\Delta\eta = 0.2$ by $\Delta\phi = 0.2$[1] RoI using two AM chips. All the results showed here are under this assumption. The set of patterns stored for an RoI is called a *pattern bank*. Each pattern bank is trained separately using a total of 30 million simulated muon tracks. The same pattern can occur more than once during training, so the patterns are sorted according to how many times they occur and only the 1 million most frequent ones are stored in the pattern bank.

In L1Track the number of patterns in the AM chips can be duplicated to speed up the matching. In EFTrack the duplication is not needed because there is no latency constraint in the Event Filter. The AM used for the duplicated pattern bank in L1Track can in EFTrack be used to expand the pattern bank to cover a p_{T} down to 2 GeV.

Ternary bits can be implemented in the AMs and used as *don't care bits*, i.e. a bit that matches regardless of whether the input is a 0 or a 1. *Don't care bits* can be used to combine patterns that are very similar to reduce the number of patterns needed. This is illustrated by track A and B in figure 5. Tracks with missing hits in one or two layers, such as track C in figure 5, can be handled using patterns with *wild card* layers, i.e. whole layers that match regardless of input.

The performance of the pattern matching has been studied for the central barrel region of $0.1 < \eta < 0.3$ and $0.3 < \phi < 0.5$ using a detailed emulation of the tracking system. The study was

[1] ϕ is the polar angle in the transverse plane of the detector and η is the pseudorapidity.

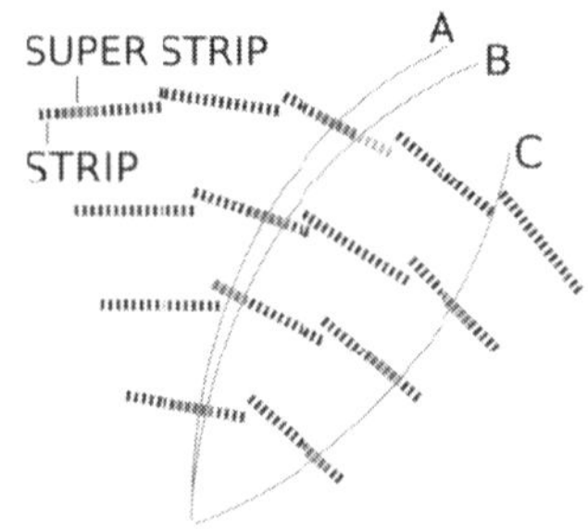

Figure 5: Patterns from tracks that only differ by one super-strip (track A and B) can be combined using *don't care bits*. Tracks with a missing hit in one layer can be handled using *wildcard layers* (track C).

performed using single muon samples with the same track parameter distributions used for training, but embedded in 200 minimum bias events. The efficiency is shown in table 2, and is close to 99.5 % for both studied cases.

3.1.2 Hough Transform

The Hough transform is a possible alternative to using AM for the pattern matching step. One advantage of the Hough transform is that it can be implemented in commercially available FPGAs and so does not require development of custom ASICs. In general, the Hough transform is used to detect features that can be described by a few parameters, such as lines and circles, in image-like data. It does so by mapping each data point to the set of all parameter values consistent with the data point and accumulating them in a histogram-like parameter space called an *accumulator*. The point in the accumulator with the largest value corresponds to a candidate feature.

In the case of tracks of charged particles in the transverse plane of the ATLAS detector, the feature we are looking for is a circular arc. If the vertex is constrained to the collision point and if the track's polar angle ϕ_0 is small, the track through a hit with the polar coordinates (r, φ) is described by the transverse momentum p_T, the charge of the particle q, and ϕ_0 by

$$\frac{A}{qp_T} = \frac{\phi_0 - \varphi}{r}, \tag{1}$$

where $A \approx 3 \times 10^{-4}\,\text{GeV}\,\text{mm}^{-1}$ is the curvature constant for the 2 T magnetic field in the ATLAS tracker. Figure 6 shows an illustration of applying equation 1 to a toy model of a tracker. The coordinates in the accumulator where the lines from the signal cross are the track parameters of that track.

An implementation of a Hough transform using equation 1 has been studied with the ATLAS ITk layout. Just as for pattern matching using AM chips, the method is looking at RoIs of size $\Delta\phi = 0.2$ by $\Delta\eta = 0.2$ separately. The accumulator is implemented as a two-dimensional histogram with $A/(qp_T)$ on one axis and ϕ_0 on the other. Each bin holds the information about which layers have been hit and the associated hits hashes for the part of the parameter space the bins spans. After the Hough accumulator has been filled, a threshold is applied to require that at least 6 unique detector layers have been hit. Nearby bins that pass the selection are grouped together and duplicate hits are removed. The end result is a collection of hit groups, i.e. hits that might be from the same track, that can be passed on to the track fitter.

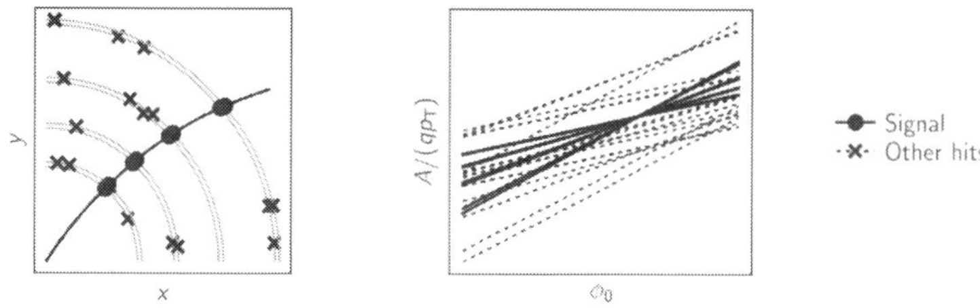

Figure 6: An illustration of the Hough transform applied to hits along a track and hits not associated with a track in the transverse plane of a tracker. Hits on the same track are mapped to crossing lines in the parameter space.

To reduce the number of hits in each bin in the accumulator, the RoI is sliced up in z, as illustrated in figure 7, and an accumulator is filled for each slice. This greatly reduces the occupancy in the $A/(qp_{\mathrm{T}})$ vs. ϕ_0 accumulator space, which in turn reduces the number of hit combinations that the track fitter needs to take into account. However, it comes at the cost of having more accumulators. There will also be an overlap between RoI slices. This makes it possible for one hit to end up in more than one hit group, which increases the number of hit combinations. In the end, it is a trade-off between the two effects.

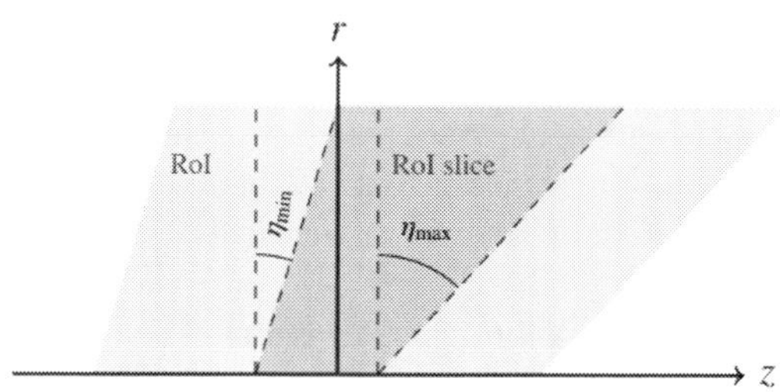

Figure 7: The RoI can be sliced up in z to reduce the occupancy in the Hough accumulator. The full RoI is given by the blue area in the figure (extending all the way from the left to the right), while a RoI slice is given by the red area. η_{min} and η_{max} are the RoI boundaries in pseudorapidity.

The efficiency of the Hough transform has been simulated in the central barrel region of $0.1 < \eta < 0.3$ and $0.3 < \phi < 0.5$. The study was performed using the same simulation samples used in studying the performance of the pattern matching using AM chips. Just as for the AM-based system, eight ITk layers were used. It was found that slicing up the RoI in approximately 12 slices in z and having an accumulator with approximately 1200 bins in $A/(qp_{\mathrm{T}})$ and 50 bins in ϕ_0 produced good results. An efficiency of 98.3 % was achieved while having a background rejection close to 80 % [8]. Although pattern matching using AM chips continues to be the main focus, the Hough transform will be continued to be studied in more detail.

3.2 Track Fitting

Track fitting is performed in an FPGA. It takes the full-resolution hits from the patterns passed by the pattern matching and calculates the track parameters and χ^2 of the fit. The track parameters p_i are calculated using a linear interpolation:

$$p_i = \sum_{j=1}^{N} C_{ij} x_j + q_i \tag{2}$$

where x_j are the full-resolution hit cluster positions and (C_{ij}, q_j) are constants that are unique for each sector (set of ITk modules). The constants are trained using simulated tracks of muons with the same track parameter ranges and distributions as used to train the patterns.

The simulated track fitting efficiency and the average number of fits for minimum bias events with pile-up 200 is shown in table 2. Two detector layer configurations are used in the simulation, one with 8 strip layers only and one where a strip layer is replaced with the outer pixel layer. The efficiency reaches above 99.5 % with respect to the events with a matched pattern.

The track parameter resolutions, defined as the RMS of the residuals with the simulated true particle, are presented in table 3 for the central barrel region. The resolutions for p_T, ϕ, η, and the impact parameter d_0 are very similar for the strip only and strip plus one pixel case while the resolution in z_0, i.e. the coordinate of the interaction point along the beam axis, is significantly improved by using one pixel layer.

Table 2: Pattern matching (using AM chips) and track fitting efficiency for simulated muons of $p_T > 4\,\text{GeV}$ in the central barrel region of $0.1 < \eta < 0.3$ and $0.3 < \phi < 0.5$. The pattern matching efficiency is defined as the fraction of single muon events with a matching pattern. The track fitting efficiency is defined as the fraction of events selected by the pattern matching that has a successful fit with $\chi^2 < 40$. Also shown is the average number of track fits needed in minimum bias events with a pile-up of 200. [9].

Detector layers	Matching eff.	Fitting eff.	$< N_{\text{fits}} >$
Strip only	99.4 %	99.5 %	114
Strip + 1 pixel	99.5 %	99.7 %	331

Table 3: Track parameter resolution for simulated muons of $p_T > 4\,\text{GeV}$ embedded in 200 minimum bias events in the central barrel region of $0.1 < \eta < 0.3$ and $0.3 < \phi < 0.5$. The resolution is defined as the RMS of the residual between the fit and the simulated true particle track parameters [9].

Layout	q/p_T [GeV^{-1}]	ϕ [rad]	η	d_0 [mm]	z_0 [mm]
Strip only	0.003	0.001	0.002	0.3	1.7
Strip + 1 pixel	0.003	0.001	0.001	0.2	0.3

3.3 Trigger Selection

The track parameters and the χ^2 computed by the track fit can be used as trigger parameters. Preliminary studies on the trigger efficiency have been performed using two simple selection strategies. The first one, strategy A, is selecting the maximum p_T track while the second one, strategy B, selects the maximum p_T track of the two tracks with best χ^2.

Figure 8 shows the background versus signal trigger efficiency for events triggered by the muon and electromagnetic triggers. The signal is represented by single muons (left) and electrons (right) as described before, while the background is simulated by muons from b-jets and jets respectively, with the same p_T distribution as expected by the L0 system (MU20 and EM18 respectively). The muon seeded trigger reaches an efficiency of almost 98 % while rejecting 80 % of the background. The electromagnetic seeded trigger reaches close to 93 % efficiency while rejecting 80 % of the background. Strategy B performs better than strategy A in both cases. By using more advanced strategies even better results are expected.

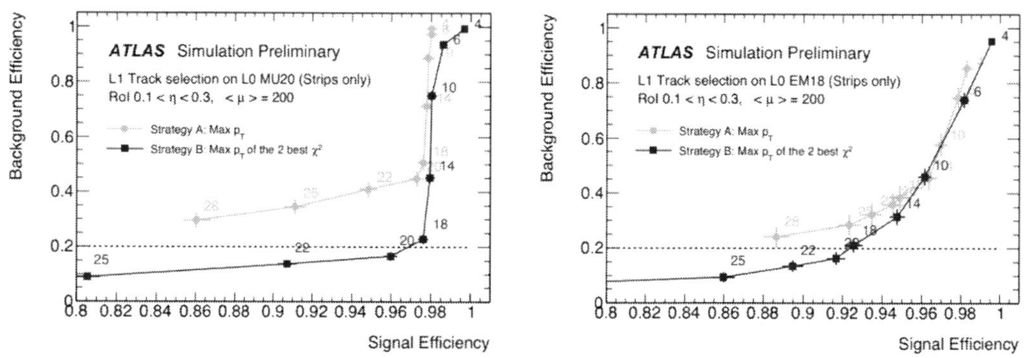

Figure 8: Signal efficiency vs. background efficiency when the track trigger is seeded by muons (left) and by the electromagnetic calorimeter trigger (right). The data point labels show the p_T cut applied in GeV [5].

4 Summary and Outlook

ATLAS will use a single-level hardware trigger, running at a maximum rate of 1 MHz, based on muon and calorimeter information. The trigger can, with increasing luminosity or if new trigger demands arise, evolve into a two-level trigger with L1Track providing regional tracking at a rate up to 4 MHz. In the single-level trigger case the regional track trigger, here called EFTrack, will assist the Event Filter. The readout latency of the tracker is below 10 μs for both trigger scenarios, which is required by the detector systems in ATLAS to avoid saturating frontend buffers.

The regional track trigger consists of a pattern matching step that selects hits followed by a track fitting step. An efficiency of 99.5 % is reached by the pattern matching step (98.3 % for the alternative method using the Hough transform) and 99.7 % for the track fitting. Preliminary studies on the track selection shows a trigger efficiency of 98 % for a muon seeded track trigger and 93 % for a calorimeter seeded track trigger with backgrounds reduced by a factor of 5.

The regional track trigger will continue to be studied and optimised based on detailed emulation of the system. Studies of the impact of different proposed ITk layouts on the trigger performance are under way, in preparation of the technical design report due to be ready by the end of 2017.

References

[1] ATLAS Collaboration, JINST **3**, S08003 (2008)

[2] ATLAS Collaboration, Tech. Rep. CERN-LHCC-2012-022. LHCC-I-023, CERN, Geneva (2012), https://cds.cern.ch/record/1502664

[3] A. Collaboration, *Phase II TDAQ upgrade physics and performance public results*, https://twiki.cern.ch/twiki/bin/view/AtlasPublic/PhysicsAndPerformancePhaseIIUpgradePublicResults, Accessed: 2017-04-25

[4] ATLAS Collaboration, Tech. Rep. CERN-LHCC-2015-020. LHCC-G-166, CERN, Geneva (2015), http://cds.cern.ch/record/2055248

[5] ATLAS Collaboration, Tech. Rep. CERN-LHCC-2017-005. ATLAS-TDR-025, CERN, Geneva (2017), https://cds.cern.ch/record/2257755

[6] A. Annovi, M. Beretta, G. Calderini, F. Crescioli, L. Frontini, V. Liberali, S.R. Shojaii, A. Stabile (ATLAS Collaboration), Tech. Rep. ATL-DAQ-PROC-2016-017, CERN, Geneva (2016), http://cds.cern.ch/record/2228284

[7] ATLAS Collaboration, Eur. Phys. J. C **70**, 823 (2010), 1005.4568

[8] M. Mårtensson (ATLAS Collaboration), Tech. Rep. ATL-DAQ-PROC-2016-034, CERN, Geneva (2016), https://cds.cern.ch/record/2234837

[9] P.O.J. Gradin (ATLAS Collaboration), Tech. Rep. ATL-DAQ-PROC-2016-013, CERN, Geneva (2016), https://cds.cern.ch/record/2216460

Track vertex reconstruction with neural networks at the first level trigger of Belle II

Sara **Neuhaus**[1,a], Sebastian **Skambraks**[1], and Christian **Kiesling**[1]

[1] *Max Planck Institute for Physics, Munich, Germany*

Abstract. The track trigger is one of the main components of the Belle II first level trigger, taking input from the Central Drift Chamber (CDC). It consists of several stages, first combining hits to track segments, followed by a 2D track finding in the transverse plane and finally a 3D track reconstruction. The results of the track trigger are the track multiplicity, the momentum vector of each track and the longitudinal displacement of the origin or production vertex of each track ("z-vertex"). The latter allows to reject background tracks from outside of the interaction region and thus to suppress a large fraction of the machine background. This contribution focuses on the track finding stage using Hough transforms and on the z-vertex reconstruction with neural networks. We describe the algorithms and show performance studies on simulated events.

1 Introduction

Belle II [1] is an upgrade of the Belle experiment, which is currently being constructed at KEK in Tsukuba, Japan. It is located at the asymmetric e^+e^- collider SuperKEKB [2], which will operate at center of mass energies at and around the $\Upsilon(4S)$ resonance. The $\Upsilon(4S)$ decays almost exclusively to $B\overline{B}$ pairs, which are nearly at rest in the center of mass system. The boost of the system allows to measure the difference between the two B decay times and thus to study time-dependent CP violation in the B meson system. SuperKEKB is designed to achieve an instantaneous luminosity of $8 \times 10^{35}\,\mathrm{cm}^{-2}\,\mathrm{s}^{-1}$ and deliver $50\,\mathrm{ab}^{-1}$ of physics events in total.

The first level trigger is a pipelined hardware trigger implemented on FPGAs. It is restricted to a maximal trigger rate of 30 kHz and a fixed total latency of at most 5 µs. The average number of charged tracks for a $B\overline{B}$ event is ≈ 10. About 3 to 9 tracks are typically visible in the track trigger, while the background track multiplicity is relatively low, producing rarely more than two visible tracks. Therefore a trigger will generally be set for three or more tracks. However, there are many interesting physics channels that cannot be efficiently triggered if three tracks are required. Examples include B decays to invisible final states and τ pair events. The latter will be produced at SuperKEKB with a cross section of the same order of magnitude as the $\Upsilon(4S)$ cross section, so τ physics are also an important aspect of the Belle II physics program.

If the trigger threshold is lowered to two tracks, additional constraints are required to suppress background. A dominant source of background comes from the machine, that is from intra-bunch

[a] e-mail: sneuhaus@mpp.mpg.de

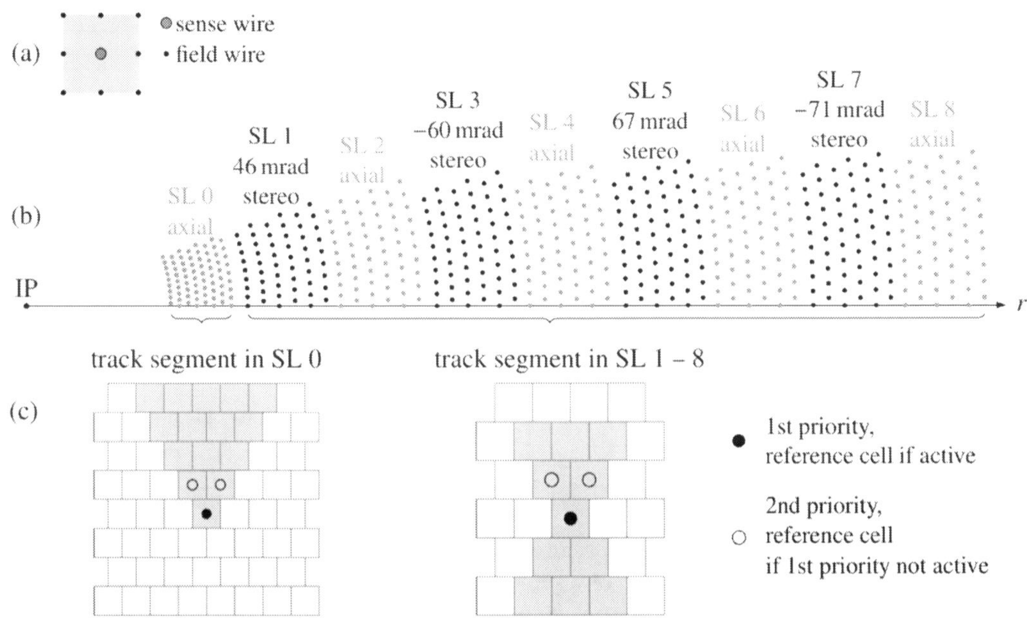

Figure 1. Wire configuration of the CDC. (a) Drift cell of one sense wire and eight field wires. (b) Arrangement of sense wires. Superlayers (SL) consist of several layers with the same wire orientation (axial or stereo). (c) In the track trigger, track segments are defined as a fixed arrangement of cells within a superlayer.

interactions (Touschek scattering [3]) and collisions with residual gas in the beampipe. This background can be suppressed by reconstructing the z-vertex of each track and rejecting tracks that do not originate at the interaction point, that is it requires a 3D track reconstruction. A second background source is Bhabha scattering, which needs to be suppressed by a factor of ≈ 100. A Bhabha veto will be based on tracks that are matched with clusters in the electromagnetic calorimeter. This matching algorithm also requires a 3D track reconstruction to obtain the polar angle and the total momentum of the track.

2 The track trigger

The input for the track trigger are hits from the Central Drift Chamber (CDC), a wire chamber with 14 336 sense wires arranged in 56 layers. The layers can be grouped into 9 so-called superlayers, which are defined by the wire orientation: 5 axial superlayers, where the wires are parallel to the z-axis, alternate with 4 stereo superlayers, where the wires are skewed by 45.4 mrad to 74.0 mrad to allow a 3D track reconstruction. Each superlayer consists of 6 layers, except for the innermost superlayer, which consists of 8 layers of smaller cells. The CDC geometry is illustrated in figure 1.

2.1 Track segment finder

The first stage of the track trigger is a data reduction step, where hits from the same superlayer are combined to so-called track segments. A track segment is defined as a fixed arrangement of wires, which is shown in figure 1c. A track segment hit is produced if there are hits in at least four of the

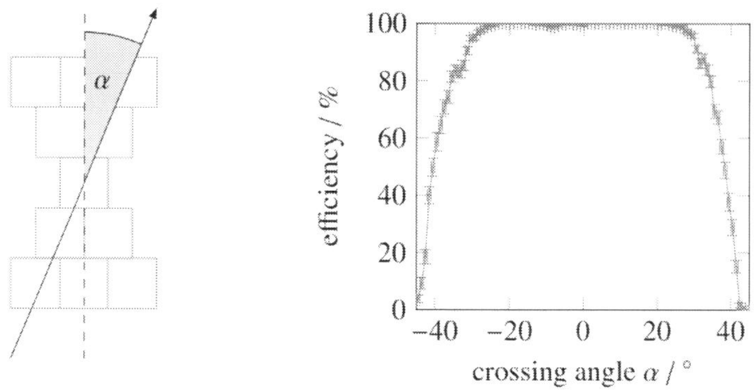

Figure 2. Track segment finder efficiency depending on the crossing angle α, measured on single track events. The efficiency is defined as the number of superlayers with a found track segment over the number of superlayers crossed by a track.

five layers within the track segment. The hit information for the following trigger stages contains a track segment number, the drift time of a reference cell in the center of the track segment, an ID that identifies the reference cell and a left/right passage. The latter tells whether the track passed left or right of the reference wire and is determined from the hit pattern within the track segment. For ambiguous hit patterns the left/right passage is undecided.

Besides compressing the input data, the track segment finder also suppresses noise, since isolated hits cannot produce a track segment. However, it also limits the acceptance of the track trigger. Figure 2 shows the track segment finder efficiency depending on the crossing angle α of the track. An efficiency of $\geq 90\,\%$ is reached for crossing angles smaller than 30°. For crossing angles larger than 45° the track segment finder is not sensitive.

2.2 2D track finder

Tracks are first searched only in the transverse plane by a Hough transform [4], which takes axial track segment hits as input. The position of the reference wire is transformed into a curve in a parameter space that is spanned by the azimuthal angle φ and the track curvature ρ, which is inversely proportional to the transverse momentum. Tracks are found as crossing points of several curves in the parameter space, with the track parameters given by the crossing point coordinates. The transformation is given by

$$\rho(\varphi) = \frac{2}{r_{\mathrm{TS}}} \cdot \sin(\varphi - \varphi_{\mathrm{TS}})$$

where $(r_{\mathrm{TS}}, \varphi_{\mathrm{TS}})$ are the polar coordinates of the reference wire. The transformation is illustrated in figure 3. Two crossing points are found for a single track, one corresponding to a track going clockwise, the other corresponding to a track going counter-clockwise. However, due to the limited track segment finder acceptance, curling tracks cannot be found in the first level trigger. Therefore, the crossing point ambiguity can be resolved by removing half of each Hough curve, namely the half that corresponds to tracks curling back through the point. The condition for the remaining outgoing half is given by

$$\varphi \in [\varphi_{\mathrm{TS}} - 90°, \varphi_{\mathrm{TS}} + 90°]$$

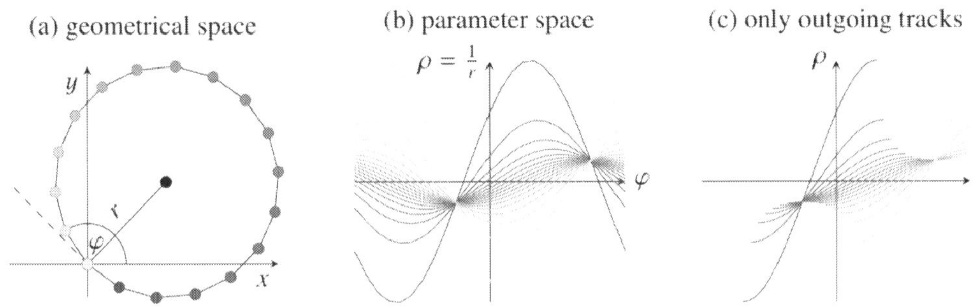

Figure 3. Illustration of the Hough transform. (a) Points on a circle in geometrical space. (b) Corresponding Hough curves in parameter space. (c) The ambiguity of the crossing point is resolved by removing the falling half of each curve.

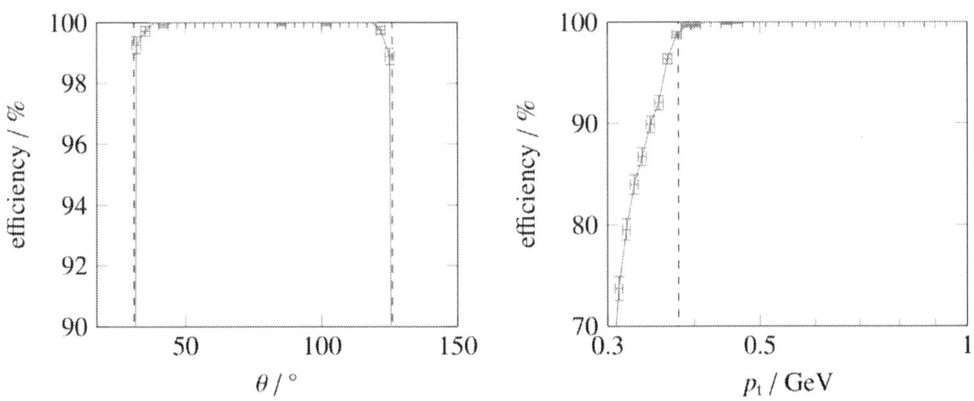

Figure 4. Track finding efficiency depending on the polar angle θ and the transverse momentum p_t, measured on single track events. Within the dashed lines the efficiency is above 99 %. For the θ efficiency, only tracks within the p_t region of full efficiency are taken into account, and vice-versa.

which corresponds to the half of the sine that has a rising slope. The crossing point of the curves for an outgoing track is then unique. The charge of the track is obtained from the crossing point as sign(ρ).

To find the crossing points, a grid of 160×34 cells is defined in the parameter space. For each grid cell, the number of crossing Hough curves from different superlayers is counted in parallel. Crossing point candidates are obtained as grid cells with at least four curves. Neighboring candidates are clustered. The track parameters φ and ρ are then given by the cluster center. The efficiency is shown in figure 4. The polar angle acceptance is given by $\theta \in [31°, 126°]$ and is directly related to the CDC geometry and the requirement of hitting at least four axial superlayers. The dependence on the transverse momentum p_t is related to the crossing angle and the track segment finder efficiency. The full efficiency is obtained for $p_t > 0.38$ GeV, which corresponds to a crossing angle of about 30° in the fourth axial superlayer. The number of grid cells was optimized to get a high track parameter resolution while maintaining a low clone rate. Clones occur when clusters have gaps, such that two clusters are found instead of one. With the selected setup, the clone rate is only 0.16 %.

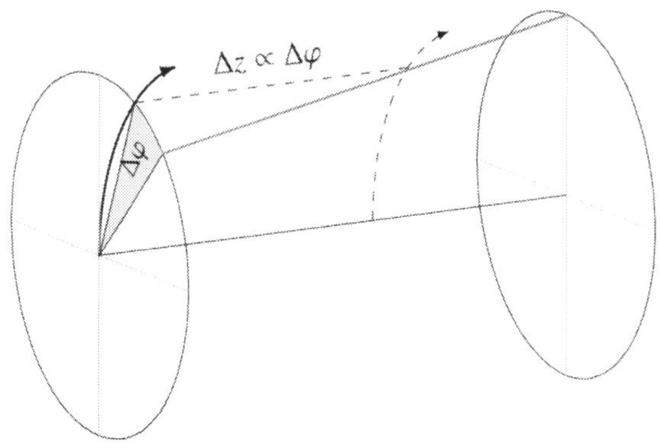

Figure 5. z-coordinates are obtained by projecting the distance $\Delta\varphi$ of a stereo wire end point from the 2D track along the stero wire.

2.3 Track vertex reconstruction

The last stage of the track trigger is a full 3D reconstruction for each track found by the Hough finder. To include the drift times in this stage, the event time needs to be known. It is determined in parallel to the track finding by comparing the "fastest timings" of all track segments, which are given by the drift time of the first hit in the track segment. Longitudinal hit coordinates can be obtained from the stereo superlayers by calculating the crossing point of a wire with the 2D track, as illustrated in figure 5. Then the longitudinal coordinates of all stereo superlayers can be combined to reconstruct the polar angle and the z-vertex of the track. The method presented here employs neural networks to estimate the 3D track parameters. Earlier versions are documented in [5–7]. In addition, an independent algorithm based on a least squares fit is being developed [8], which will run in parallel to the neural network. The optimal way to combine the results of the two algorithms for the final trigger decision has yet to be studied.

For the input of the neural network, one hit is selected for each superlayer. Hit candidates have to be within a certain φ region around the 2D track and within a certain time window. If there are several hit candidates, hits with known left/right passage and the shortest drift time are selected. Then three input values are calculated for each hit: the distance $\Delta\varphi$ between the 2D track and the wire end point, the drift time with a sign given by the left/right passage and the crossing angle α of the 2D track, which contains information about the track curvature. Using a separate input for the drift time allows the network to learn nonlinear corrections for the drift velocity and to take the crossing angle into account. Both axial and stereo hits are used. Although axial hits contain no 3D information, they allow to improve the 2D track estimate, which was obtained without drift times.

In total 27 input values are obtained, which are scaled to a common interval of $[-1, 1]$ and fed into a Multi Layer Perceptron (MLP) [9] with a single hidden layer of 127 nodes and two output nodes, as shown in figure 6. The three layer feed-forward structure is particularly suited for parallel execution with a deterministic runtime, as required by the trigger. The size of the hidden layer was optimized such that a further increase does not significantly improve the resolution of the output values. Each hidden and output neuron computes a weighted sum of its inputs and and evaluates it with an activation function:

$$y_j = \tanh\left(\sum_{i=0}^{N} x_i \cdot w_{ij}\right)$$

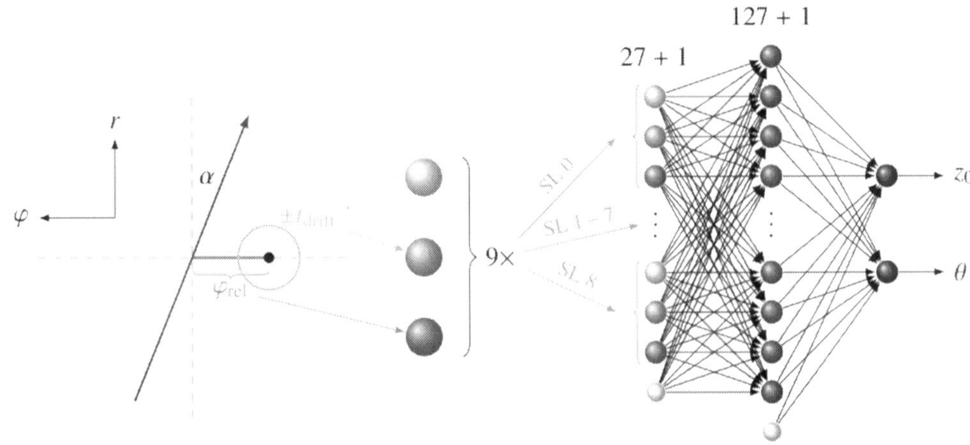

Figure 6. Illustration of the input and the structure of the MLP. The small white nodes denote bias nodes with a fixed input of 1.

Here, the x_i are the N input values for the neuron with output y_j, w_{ij} are the weights and the first weight w_{0j} acts as a constant bias, with $x_0 = 1$. The weights are trained using the backpropagation algorithm iRprop$^-$ [10] as implemented in the open source library FANN [11], with the z-vertex and the polar angle θ as continuous targets. To match the co-domain of the activation function, the target values are scaled to the interval $[-1, 1]$. To obtain the track parameters with the correct units, the inverse scaling is applied to the network outputs. Note that the scaling procedure requires a cut-off during the training, using only tracks with $|z| \leq 50$ cm.

For superlayers where no hits pass the hit selection criteria, the corresponding input values are missing. For axial layers, a default input tuple is defined as $(0, 0, 0)$, which can be interpreted as "no 2D correction for this superlayer". For stereo layers, no reasonable default values exist, as any input would be interpreted as some z-coordinate. Therefore, independent networks are trained for the case of missing stereo hits, which receive only 24 input nodes. In total five such "expert" networks are trained, one for 4 stereo hits and four for 3 stereo hits, depending on which hit is missing. For tracks with less than 3 stereo hits no 3D reconstruction is possible, which happens for ≈ 1 % of the tracks.

3 Performance

For single tracks with a momentum distribution uniform in the solid angle and the track curvature, an average resolution of 2.9 cm for the z-vertex and 3.1° for the polar angle is obtained. The resolution depends on the transverse momentum, with a z-vertex resolution of 2.3 cm for high momentum tracks and 4.3 cm for momenta at the edge of the track finding acceptance. In the following, a cut is applied to the estimated z-vertices of physics and background tracks, to demonstrate the rejection power and efficiency of a z-vertex trigger.

3.1 Background rejection

For the machine background simulation, particles are generated whose energy and momentum deviate slightly from the design values, with the exact distribution depending on the background type. Then

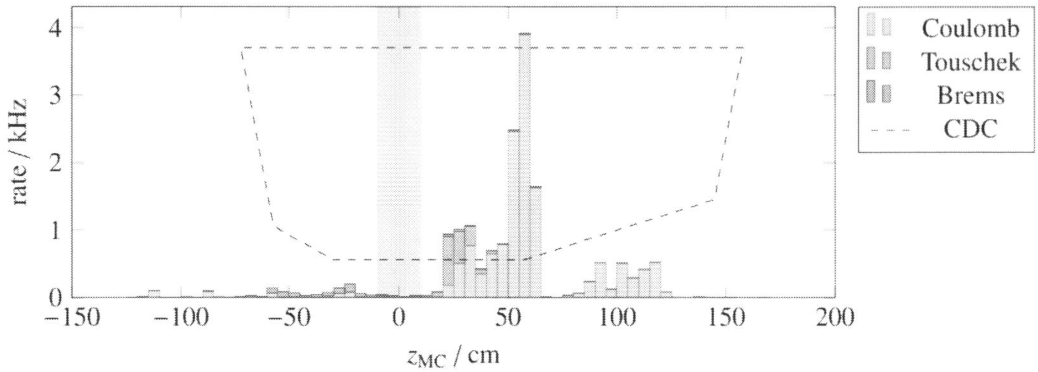

Figure 7. True z-vertex distribution of background tracks found by the trigger. The outline of the CDC is shown as a scale reference. The shaded region shows ±10 cm around the interaction point.

the particles are tracked through the accelerator ring. When particles deviate far enough from the design orbit that they hit the beampipe or other machine elements, the loss position and momentum are recorded. The initial energy and momentum deviations are simulated for three types of reactions: Touschek scattering within the bunches, elastic scattering with residual gas ("Coulomb") and inelastic scattering ("Brems"). Then a full detector simulation is done for particles that are lost in the vicinity of the detector. Figure 7 shows the distribution of true z-vertices for tracks found in the trigger. Mostly, those tracks correspond to secondary particles that are created when a lost particle is scattered in some detector structure. The integrated background track rate without a z-vertex veto is 17 kHz, which is more than half of the maximal total trigger rate. By rejecting tracks whose estimated z-vertex is outside of ±10 cm, the background rate can be reduced to 0.2 kHz. Note that a very efficient rejection is achieved also for tracks with $|z| > 50$ cm, which were not included in the training.

3.2 Track trigger efficiency

The efficiency is measured both for generic $B\overline{B}$ events and for $\tau^+\tau^-$ events, which are studied as an example for events with low multiplicity. In addition to generic τ decays, a specific channel is investigated, namely the flavor violating decay $\tau \to \mu\gamma$. Table 1 summarizes the track trigger efficiency for different trigger conditions. For three or more tracks no additional constraint is imposed. For events with two tracks, different z-vertex constraints are compared: either both tracks are required to come from the interaction point (IP) or at least one of them, where an IP track is characterized by an estimated z-vertex within ±10 cm. For comparison, the pure two track efficiency is also shown. For two IP tracks, the efficiency is relatively low compared to a pure two track trigger. The losses are due to tracks with low momentum, missing hits or wrong hit selection. Opening the z-vertex cut reduces these losses only slightly and increases the background rate. A better trade-off is obtained by requiring only one of the two tracks to originate at the interaction point. Then the efficiency is very close to a pure two track trigger.

Table 1 also shows the importance of a low multiplicity trigger for τ physics studies. Compared to a three track trigger, the efficiency is increased by a factor of 3 for generic τ events and a factor of 4.6 for the flavor violating decay mode.

Table 1. Efficiency for example event types under different trigger conditions. A certain total number of tracks is required as well as a certain number of tracks with a z-vertex within ± 10 cm ("IP tracks"). The last line shows the combined efficiency of ≥ 3 tracks or 2 tracks with IP constraint. The efficiency is calculated relative to the total number of simulated events.

trigger condition		efficiency		
# tracks	# IP tracks	$B\overline{B}$	$\tau^+\tau^-$	$\tau \to \mu\gamma$
≥ 3	≥ 0	93.9 %	18.7 %	12.4 %
2	≥ 0	4.6 %	38.7 %	44.4 %
2	≥ 1	4.4 %	38.1 %	44.1 %
2	2	2.9 %	33.5 %	40.0 %
3 or 2 with ≥ 1 IP		98.3 %	56.8 %	56.5 %

4 Conclusion

The first level track trigger of the Belle II experiment will offer a full 3D reconstruction. In particular, the z-vertex reconstruction allows to suppress the expected machine background by two orders of magnitude. This reduction is essential to open the trigger for low multiplicity events. For a hypothetical decay channel $\tau \to \mu\gamma$ an efficiency of 56.5 % is obtained for a two track trigger. This efficiency will be further increased to about 90 % by including other subtriggers, such as the calorimeter trigger and a combined track-calorimeter logic.

The neural network logic has been implemented on FPGA and verified on simulated events [12]. The neural network trigger is ready to be installed and will be tested on cosmic rays during the summer of 2017.

Acknowledgements

This work has been supported by the German Federal Ministry of Education and Research (BMBF) under grant number 05H15WOKBA. The machine background simulations have been carried out on the computing facilities of the Computational Center for Particle and Astrophysics (C2PAP) within the Excellence Cluster Universe.

References

[1] T. Abe, I. Adachi, K. Adamczyk, S. Ahn, H. Aihara, K. Akai, M. Aloi, L. Andricek, K. Aoki, Y. Arai et al. (Belle II) (2010), Belle II Technical Design Report, `http://arxiv.org/abs/1011.0352`

[2] Y. Ohnishi, T. Abe, T. Adachi, K. Akai, Y. Arimoto, K. Ebihara, K. Egawa, J. Flanagan, H. Fukuma, Y. Funakoshi et al., Progress of Theoretical and Experimental Physics **2013**, 03A011 (2013)

[3] C. Bernardini, G.F. Corazza, G. Di Giugno, G. Ghigo, J. Haissinski, P. Marin, R. Querzoli, B. Touschek, Phys. Rev. Lett. **10**, 407 (1963)

[4] R.O. Duda, P.E. Hart, Commun. ACM **15**, 11 (1972)

[5] S. Skambraks, Diploma thesis, Ludwig-Maximilians-Universität, München (2013), `https://publications.mppmu.mpg.de/2013/MPP-2013-377/FullText.pdf`

[6] S. Skambraks, F. Abudinen, Y. Chen, M. Feindt, R. Frühwirth, M. Heck, C. Kiesling, A. Knoll, S. Neuhaus, S. Paul et al., Nuclear Science, IEEE Transactions on **62**, 1732 (2015)

[7] S. Neuhaus, S. Skambraks, F. Abudinen, Y. Chen, M. Feindt, R. Frühwirth, M. Heck, C. Kiesling, A. Knoll, S. Paul et al., Journal of Physics: Conference Series **608**, 012052 (2015)

[8] J.B. Kim, Master thesis, Korea University (2012), `http://dcollection.korea.ac.kr/jsp/common/DcLoOrgPer.jsp?sItemId=000000033908`

[9] R. Hecht-Nielsen, *Theory of the backpropagation neural network*, in *1989 IJCNN, International Joint Conference on Neural Networks* (1989), pp. 593–605

[10] C. Igel, M. Hüsken, *Improving the Rprop Learning Algorithm*, in *Proceedings of the Second International Symposium on Neural Computation* (2000), pp. 115 – 121

[11] S. Nissen, Report, Department of Computer Science University of Copenhagen (DIKU) **31**, 29 (2003), code available at `https://github.com/libfann/fann`

[12] S. Bähr, S. Skambraks, S. Neuhaus, C. Kiesling, J. Becker, *A neural network on FPGAs for the z-vertex track trigger in Belle II*, in *Topical Workshop on Electronics for Particle Physics (TWEPP) 2016* (2017), Vol. 12, p. C03065

Bivariate normal distribution for finding inliers in Hough space for a Time Projection Chamber

Amir **Noori Shirazi**[1,a] and Ivor **Fleck**[1,b]

[1] *Department Physik, Universität Siegen, Walter-Flex-Str.3, D57068, Siegen, Germany*

Abstract. A Time Projection Chamber (TPC) is foreseen as the main tracking detector for the International Large Detector (ILD), one of the two detectors for the next candidate collider named International Linear Collider (ILC) [1].
GridPix, a combination of a micropattern gaseous detector with a pixelized readout, is one of the candidate readout systems for the TPC [2] [3]. One of the challenges in the track reconstruction is the large number of individual hits along a track (around 100 per cm). Due to the small pixel size of 55 x 55 μm^2, the distance between the individual ionization processes in the gas is larger than the size of the pixels. Consequently, the hits in the GridPix are not contiguous. This leads to the challenge of assigning the individual hits to the correct track. Hits within a given distance from a reconstructed track are called inliers. Consequently, finding inliers within the large number of hits and in the presence of noise is difficult for pattern recognition. This difficulty is increased by diffusion effects in the TPC.
One of the current algorithms which are utilized for track finding is the Hough transform. Using a bivariate normal distribution based on the covariance matrix calculated from the diffusion effect improves collecting inliers in the Hough space directly [5].

1 Introduction

The International Linear Collider (ILC) to be built in Japan, is one of the main candidate colliders for the future of high energy physics. The ILC has two main detectors: International Large Detector (ILD) and Silicon Detector (SiD). The main tracking detector for ILD is a Time Projection Chamber (TPC) with benefits such as low material budget and continuous tracking [1].

GridPix, one of the candidate readout systems for the TPC, is a combination of a micropattern gaseous detector with a TimePix chip readout system [2] [3]. The result of this combination is that there are many individual hits along a track (around 100 per cm). The size of each pixel is 55 x 55 μm^2 and in each chip, there are 256 x 256 pixels. Every 8 GridPix chips are aligned in a structure called Octoboard [2]. One of the main reasons to use GridPix is to reach the highest possible double track resolution for the ILD-TPC. A large number of hits makes it very challenging for the pattern recognition to assign a hit to the correct track. This difficulty is increased because of noise and diffusion effects in the TPC.

[a]e-mail: shirazi@hep.physik.uni-siegen.de
[b]e-mail: fleck@hep.physik.uni-siegen.de

The strategy that is used here is to find inliers for a small segment of a track, called tracklet, and then to connect all relevant tracklets to each other in order to have the full track. One of the advantages of this method is, that finding inliers locally is easier than using a global method for the full track. For this study, the unit of the Octoboard is for finding a segment of the track. The size of each Octoboard is 45 x 65 mm^2 and this allows for high momentum particles, to assume a straight line for each segment even with a magnetic field in the range of 4 T. It is obvious that this assumption is not correct for low momentum particles, for instance in the range of a few MeV.

Hough transform is one of the typical methods used in image processing, computer vision and also in high energy physics to find the parameters of a line or curve fitted on many points or hits in an image space based on a voting process. According to Richard Duda and Peter Hart, the normal equation of a straight line becomes [4]:

$$\rho = x\cos(\phi) + y\sin(\phi) \tag{1}$$

Where ρ and ϕ are the normal parameters of the straight line. Based on equation (1) each hit generates a sinusoid shaped line in Hough space and the lines of all hits intersect in a bin that corresponds to a parameter vector. In order to find the track parameters, the bin with the maximum entries in Hough space is taken and this bin returns the candidate parameters of a candidate track [6]. As mentioned before, due to the diffusion effects in the TPC, the width of a track increases with the drift distance and for a fixed size of the bins in Hough space, some lines for inliers do not pass through the bin corresponding to the track parameters.

This paper explains how to use a bivariate normal distribution (BND) fit in Hough Space around the bin with the maximum entries in order to collect inliers in Hough space [5]. With this method, we are able to separate two tracks from each other and also to separate track hits from noise hits. In the next section we explain how to calculate BND fit parameters based on the diffusion coefficient.

2 Diffusion and Covariance matrix

After the primary ionization created by a charged particle along its trajectory inside a TPC, a charge cloud of electrons from ionization in a gas diffuses because of the thermal movement of the gas molecules. The diffusion effect increases with the drift distance and causes the charge cloud, after some time t to show a Gaussian density distribution as follows [7] [8]

$$n = \frac{1}{\sqrt{4\pi\tilde{D}_L t}}\left(\frac{1}{\sqrt{4\pi\tilde{D}_T t}}\right)^2 \exp\left(-\frac{x^2+y^2}{4\tilde{D}_T t} - \frac{(z-v_d t)^2}{4\tilde{D}_L t}\right) \tag{2}$$

Where v_d is the drift velocity and $\tilde{D_T}$ and $\tilde{D_L}$ are transverse and longitudinal diffusion coefficient respectively. When applying an electric field for drifting, the diffusion in the Z direction (along the electric field) is different from the X and Y directions which are perpendicular to the electric field. For this reason, there are two diffusion coefficients, one for the X and Y directions ($\tilde{D_T}$) and another one for the Z direction ($\tilde{D_L}$). According to equation (2) there are two resolutions as follows:

$$\sigma_T = \sqrt{2\tilde{D}_T t} \quad \text{and} \quad \sigma_L = \sqrt{2\tilde{D}_L t} \tag{3}$$

Where σ_T and σ_L are transverse and longitudinal resolution respectively. In order to eliminate the dependency on the time, diffusion constants are defined as $D_T = \sqrt{\frac{2\tilde{D}_T}{v_d}}$ and $D_L = \sqrt{\frac{2\tilde{D}_L}{v_d}}$ for both diffusion coefficients. The diffusion constants have a dependency on the gas mixture used in the TPC.

With implementing a magnetic field parallel to the electric field, the trajectory of the electrons has a helix shape and it causes the transverse diffusion constant to be different from that without magnetic field. In addition to the diffusion the resolutions also depend on the readout system. The transverse resolution has a contribution $\sigma_0 = 0.055\ \text{mm}/\sqrt{12}$ from the pitch of the pixels and the longitudinal resolution a contribution due to the frequency (f_c) of the readout system. In our case $f_c = 40$ MHz [9]. In this case, the transverse and longitudinal resolution have the form:

$$\sigma_T = \sqrt{\sigma_0^2 + Z_{drift} D_T^2}$$
$$\sigma_L = \sqrt{\frac{1}{6}\left(\frac{v_d}{f_c}\right)^2 + Z_{drift} D_L^2} \tag{4}$$

Which shows the resolution of a track in dependence on the drift length, the gas mixture and the magnetic field [8] [9].

Generally, for GridPix as the readout system, because of the small size of the pixels, only a single electron is detected in one readout channel. Therefore there is no correlation between X and Y coordinates. Consequently, the covariance matrix is a 3x3 diagonal matrix based on the transverse and longitudinal diffusion.

$$cov = \begin{bmatrix} \sigma_T^2 & 0 & 0 \\ 0 & \sigma_T^2 & 0 \\ 0 & 0 & \sigma_L^2 \end{bmatrix} \tag{5}$$

For the ideal case, when a fixed size of the binning is used for Hough space, and the width of the charge cloud is small, the lines of all hits pass through one bin. This bin then corresponds to track parameters in Hough space. But when the diffusion is large, then for the same binning, some lines could not pass through the bin with maximum entries. For this reason, in order to find inliers directly inside the Hough space, we calculate the resolution of the track parameters in this bin. The following section explains how it is done. Figure 1 is an example for the image and Hough space for a track.

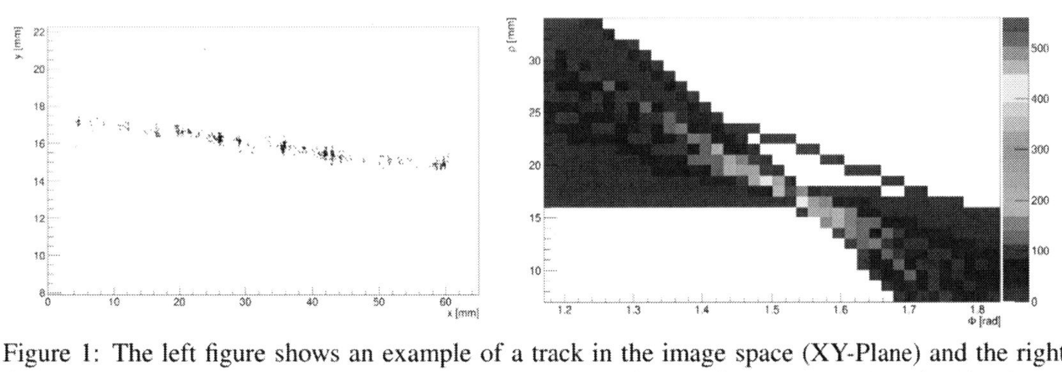

Figure 1: The left figure shows an example of a track in the image space (XY-Plane) and the right figure shows the Hough space for the same hits. Because of the diffusion, some lines for hits do not pass through the bin with maximum entries.

DOI: 10.1051/epjconf/201715000010

2.1 Covariance matrix in the XY-Plane

As mentioned before, we use the normal equation for Hough transform. For the first step we calculate the track's parameters only in the XY-Plane. In this case, the covariance matrix for image space (or for hits space) is

$$V_{ij} = \begin{bmatrix} \sigma_T^2 & 0 \\ 0 & \sigma_T^2 \end{bmatrix} \tag{6}$$

The next step is to calculate the covariance matrix in Hough space. For this calculation, we use first-order error propagation:

$$\begin{bmatrix} \sigma_\rho^2 & \sigma_{\rho\phi} \\ \sigma_{\rho\phi} & \sigma_\phi^2 \end{bmatrix} = J \begin{bmatrix} \sigma_T^2 & 0 \\ 0 & \sigma_T^2 \end{bmatrix} J^T \tag{7}$$

With $\sigma_{\rho\phi}$, σ_ρ and σ_ϕ are covariance and variance of ρ and ϕ in Hough Space, respectively. In addition, J is the Jacobian matrix of equation (1) [10].

$$J = \begin{bmatrix} \frac{\partial\rho}{\partial x} & \frac{\partial\rho}{\partial y} \\ \frac{\partial\phi}{\partial x} & \frac{\partial\phi}{\partial y} \end{bmatrix} \tag{8}$$

2.2 Covariance matrix in the SZ-Plane

In order to separate two tracks or better say to find inliers in the Z direction, the Hough transform is also implemented in the SZ-Plane. S is defined as the arc length in the XY-Plane, and in the case of B=0 (or straight tracks), is the shortest distance between a hit and the point of closest approach (pca). The pca is the closest point on a track to the reference point (here: the origin of the coordinate system) [11].

$$S = \sqrt{(x_{hit} - x_{pca})^2 + (y_{hit} - y_{pca})^2} \tag{9}$$

For calculating σ_S first order error propagation is used based on equation (9)

$$\sigma_S^2 = \left(\frac{\partial S}{\partial x}\right)^2 \sigma_T^2 + \left(\frac{\partial S}{\partial y}\right)^2 \sigma_T^2 \tag{10}$$

In the SZ-Plane, the covariance matrix is defined as:

$$V_{ij} = \begin{bmatrix} \sigma_S^2 & 0 \\ 0 & \sigma_L^2 \end{bmatrix} \tag{11}$$

For the Hough transform in the SZ-Plane we use:

$$\rho_z = S\cos(\theta) + z\sin(\theta) \tag{12}$$

And therefore we have

$$\begin{bmatrix} \sigma_{\rho_z}^2 & \sigma_{\rho_z\theta} \\ \sigma_{\rho_z\theta} & \sigma_\theta^2 \end{bmatrix} = J \begin{bmatrix} \sigma_S^2 & 0 \\ 0 & \sigma_L^2 \end{bmatrix} J^T \tag{13}$$

With J is, as before, the Jacobian matrix [10]:

$$J = \begin{bmatrix} \frac{\partial\rho_z}{\partial S} & \frac{\partial\rho_z}{\partial z} \\ \frac{\partial\theta}{\partial S} & \frac{\partial\theta}{\partial z} \end{bmatrix} \tag{14}$$

3 Bivariate normal distribution (BND)

In order to find inliers and outliers in Hough space, bivariate normal distribution is utilized around the bin with maximum entries in Hough space. Generally, a bivariate normal distribution shows the distribution of two parameters which could have a correlation. In Hough space, two parameters (for 2D) ϕ and ρ have a correlation. The general equation for the bivariate normal distribution is:

$$G(\phi,\rho) = \frac{1}{2\pi\sigma_\rho\sigma_\phi\sqrt{1-r^2}}\exp\Big(-\frac{c}{2(1-r^2)}\Big) \tag{15}$$

where

$$c \equiv \frac{(\phi-\mu_\phi)^2}{\sigma_\phi^2} - \frac{2r(\phi-\mu_\phi)(\rho-\mu_\rho)}{\sigma_\phi\sigma_\rho} + \frac{(\rho-\mu_\rho)^2}{\sigma_\rho^2} \tag{16}$$

and μ_ϕ and μ_ρ are the mean value for ϕ and ρ respectively and r is the correlation

$$r \equiv \frac{\sigma_{\rho\phi}}{\sigma_\rho\sigma_\phi} \tag{17}$$

The Bivariate normal distribution has an elliptical shape. In case there is a correlation between the two parameters, this ellipse has a tilt to the x-axis otherwise it would be a vertical or horizontal ellipse (depends on the minor and major semi-axis). In Eq (15), the term before the exponential part is the normalization factor [13].

After fitting the BND in Hough space around the bin with maximum entries using ROOT [12], there is an iso-contour of equation (15) which has an elliptical shape and then one can find the semi-minor and semi-major axis of the ellipse corresponding to the 1/e of the peak of the BND fit. This ellipse corresponds to the one sigma interval of the fit. A hit that its line in Hough space passes through this ellipse is called an inlier. The important point of this method is the size of the ellipse is changeable which means the number of inliers can be changed. Figure 2 shows the BND fit in Hough space (a) and the ellipse corresponding to the 1σ interval of the (b).

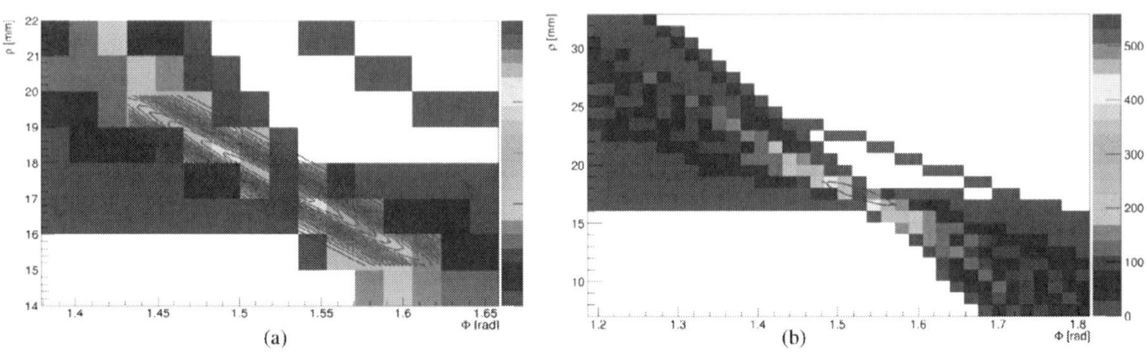

Figure 2: (a) Iso-countor of the BND fit and (b) the ellipse corresponding to 1/e of the peak value.

From equation(15) and (16) we have

$$1 = \frac{(\phi-\mu_\phi)^2}{q\sigma_\phi^2} + \frac{(\rho-\mu_\rho)^2}{q\sigma_\rho^2} - \frac{2r(\phi-\mu_\phi)(\rho-\mu_\rho)}{q\sigma_\phi\sigma_\rho} \tag{18}$$

and also generally the equation of a tilted ellipse is

$$1 = \left(\frac{\cos^2(\alpha)}{a^2} + \frac{\sin^2(\alpha)}{b^2}\right)\phi^2 - 2\cos(\alpha)\sin(\alpha)\left(\frac{1}{a^2} - \frac{1}{b^2}\right)\phi\rho + \left(\frac{\sin^2(\alpha)}{a^2} + \frac{\cos^2(\alpha)}{b^2}\right)\rho^2 \quad (19)$$

therefore we have semi-major and semi minor of the ellipse as follow:

$$a^2 = \frac{q\sigma_\phi^2\sigma_\rho^2\cos(2\alpha)}{\sigma_\rho^2\cos^2(\alpha) - \sigma_\phi^2\sin^2(\alpha)}$$
$$b^2 = \frac{-q\sigma_\phi^2\sigma_\rho^2\cos(2\alpha)}{\sigma_\rho^2\sin^2(\alpha) - \sigma_\phi^2\cos^2(\alpha)} \quad (20)$$

After finding the two ellipses for both, the XY-plane and the SZ-plane, we iterate over all hits and those having a line that passes through both ellipses are collected as inliers. Thereafter, a Karimäki Circle fit and a straight line fit are used for having pre-estimations of the parameters of the tracklet which later will be used for the merging process [14] [15].

4 Result

As mentioned before, one of the objectives of our study is to reach the highest possible double track resolution. As the first step, we simulated two electrons with an energy of 6 GeV to pass through the TPC parallel to the x-axis. All simulations were done for the Large Prototype Time Projection Chamber (LP-TPC) at DESY with a 1T magnetic field. These simulations were done for two tracks with the same Z position and varying distances in Y (1-8 mm) (figure 3) and then for the same condition but this time for the variation of the distance between 1 mm to 10 mm in the Z direction and the same Y offset (figure 4). All simulations were done near the full drift length of 600 mm in order to have the maximum diffusion effect. For each step 1000 events were generated, with up to 500 additional random noise hits for each event.

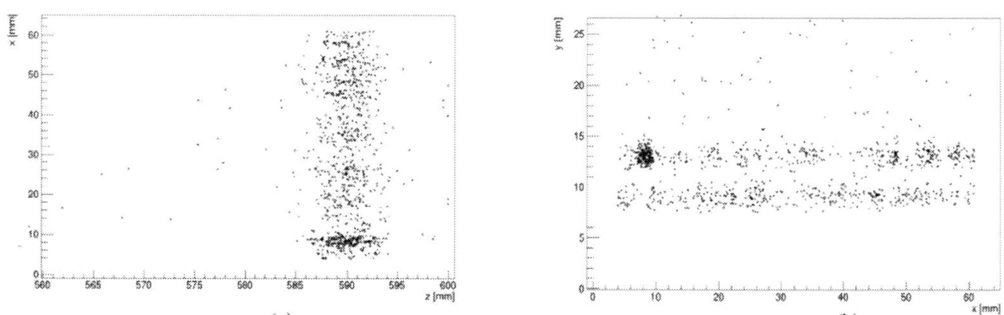

(a) (b)

Figure 3: Example for two tracks with the same Z offset and $\Delta Y = 4$mm.(a) The two tracks in the XZ-Plane, and (b) the same two tracks in the XY-Plane.

DOI: 10.1051/epjconf/201715000010

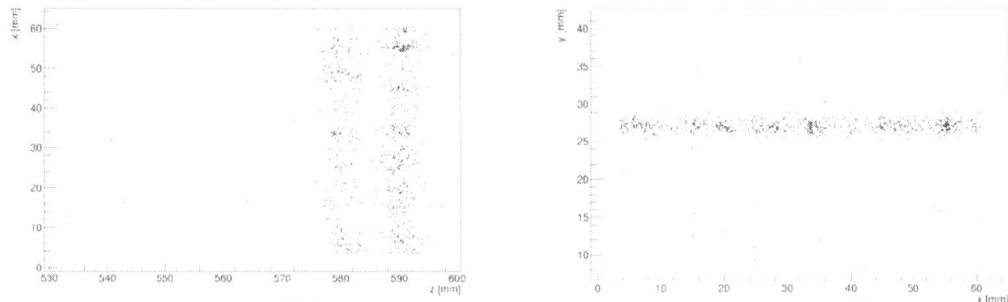

(a) (b)

Figure 4: Example for two tracks with the same Y offset and $\Delta Z = 4$mm.(a) The two tracks in the XZ-Plane, and (b) the same two tracks in the XY-Plane.

The results of the reconstruction for different track distances are shown in figure 5. To calculate the track efficiency, a correct track is associated with its true Monte Carlo particle if it has more than some percentage (e.g. 60 % or 70 %) of that MC particle's hits otherwise it is identified as a ghost track. In the case that a reconstructed track has sufficient hits for both MC particles, it is assigned to only one of them.

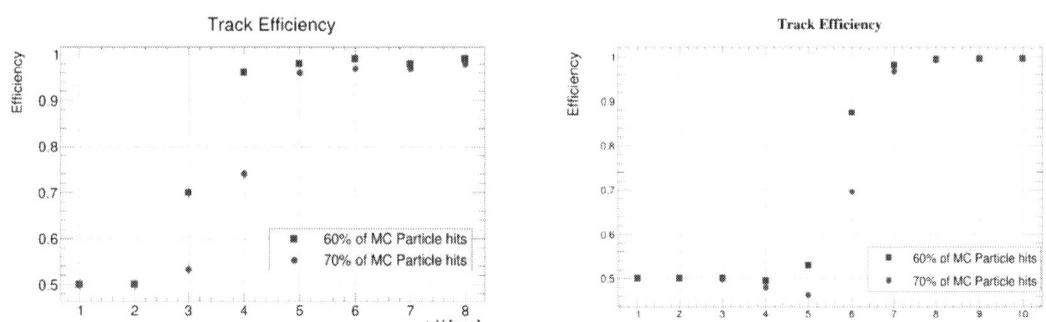

(a) (b)

Figure 5: Track efficiency for tracks having more than 60 % (blue points) and 70 % (red points) of the true MC particle's hits.(a) Track efficiency as a function of ΔY. (b) Track efficiency as a function of ΔZ.

For these simulations the transverse resolution (σ_T) was about 0.7 mm and the longitudinal resolution (σ_L) was approximately 2.0 mm. According to the plot for ΔY simulation, for correct tracks with more than 60 % of MC particle's hits, we are able to separate double tracks with 4 mm distance from each other for more than 90 % of the simulated tracks. However, the efficiency of this simulation when we consider correct tracks with 70 % of correct MC particle hits decreases to around 75 %. For distance of 2 mm or less the two tracks cannot be separated any more and only one correct track can be reconstructed. This results in a track efficiency of 50 %. We can see the same behaviour for the ΔZ simulation plot for 6 mm distance between two tracks.

5 Conclusion

In summary, we used a bivariate normal distribution fit in Hough space based on the covariance matrix calculated according to the transverse (σ_T) and longitudinal (σ_L) resolution in order to collect inliers directly in Hough space. The first study of this algorithm shows this method is promising. In the case that a correct track is associated with its true MC particle on condition that it has more than 60 % of MC particle's hits, the simulation of two tracks for near the full drift length (600 mm) for LP-TPC shows that we can reach to more than 90 % track efficiency when the distance of two tracks is 4 mm in the Y direction and have the same Z offset. This result is also true for the case that two tracks separated from each other with 6 mm distance in the Z direction and the same Y offset. The transverse resolution for this analysis and based on the equation (4) is about 0.7 mm and the longitudinal resolution is 1.8 mm.

According to the LC-TPC characteristics, the double track resolution in the $r\phi$ plane should be around 2 mm and in the rz plane should be almost 6 mm for the full drift length (2.35 m) of the TPC with a 3.5 T magnetic field [1]. In such a case, with the same electric field implemented in our simulation, the transverse resolution is about 0.4 mm and the longitudinal resolution is around 3.6 mm for the full drift length of the LC-TPC. As a conclusion, with extrapolation of our results to the LC-TPC conditions, we expect that this method will be able to achieve the double track resolution required for the LC-TPC.

In addition, comparing the longitudinal resolution required for the LC-TPC with that in this simulation shows that to reach the double track resolution in the rz plane could be an issue, however, more studies and optimization of this method are ongoing. Different parameters are tuned to find any possibility of having better results and to improve this algorithm. The next step in finalizing the pattern recognition for a pixel readout TPC, will be to develop an algorithm for correctly merging the tracklets found with the method described here.

References

[1] *The International Linear Collider Technical Design Report - Volume 4: Detectors*, 2013. CERN-ATS-2013-037, DESY 13-062, ILC-REPORT-2013-040.

[2] Michael Lupberger. *The Pixel-TPC: A feasibility study*. PhD thesis in Physik Bonn University, Germany, Bonn, August, 2015.

[3] X. Llopart et al. *Timepix, a 65k programmable pixel readout chip for arrival time, energy and/or photon counting measurements.*, Nuclear Instruments and Methods in Physics Research Section A: Accelerators, Spectrometers, Detectors and Associated Equipment 581.1 (2007) 485–494, doi: 10.1016/j.nima.2007.08.079 (cit. on pp. 25, 30, 34, 112).

[4] Richard O.Duda and Peter E .Hart. *Use of the Hough Transformation To Detect Lines and Curves in Pictures*. Standford Research Institute, Menlo Park, California,1972.

[5] Leandro A.F.Fernandes, Manuel M .Oliveira. *Real -time line detection through an improved Hough transform voting scheme*. Elsevier,Pattern Recognition 41 (2008) 299 – 314.

[6] Richard O.Duda and Peter E .Hart. *Use of the Hough Transformation To Detect Lines and Curves in Pictures*. Standford Research Institute, Menlo Park, California,1972.

[7] W. Blum and L. Rolandi. *Particle Detection with Drift Chambers*. Springer, Berlin, 1993.

[8] Christoph Brezina. *A GEM Based Time Projection Chamber with Pixel Readout*. zur Erlangung des Doktorgrades (Dr. rer. nat.) der Mathematisch-Naturwissenschaftlichen Fakultät, der Rheinischen Friedrich-Wilhelms-Universität Bonn, 2013

[9] Martin Rogowski. *Implementation of a track finding algorithm for an InGrid based detector*. Masterarbeit in Physik Bonn University, Germany, Bonn, November, 2014.

[10] Kai Oliver Arras. *An Introduction To Error Propagation: Derivation, Meaning and Examples of Equation* $C_Y = F_X C_X F_X^T$. Technical Report N° EPFL-ASL-TR-98-01 R3 of the Autonomous Systems Lab, Institute of Robotic Systems, Swiss Federal Institute of Technology Lausanne (EPFL),September 1998.

[11] Isa Heinze. *Development of a Hough Transformation Track Finder for Time Projection Chambers*. zur Erlangung des Doktorgrades, des Fachbereichs Physik, der Universität Hamburg, Hamburg, 2013.

[12] Rene Brun and Fons Rademakers. Root - an object oriented data analysis framework. Nucl. Instrum. Meth. A, 389:81–86, 1997.

[13] Glen Cowan. *Statistical Data Analysis*. Oxford University Press Inc., New York, 1998.

[14] Thomas Krämer. *Track parameters in LCIO*. 2006. LC-DET-2006-004.

[15] V. Karimäki. *Effective circle fitting for particle trajectories*. Nuclear Instruments and Methods in Physics Research Section A: Accelerators, Spectrometers, Detectors and Associated Equipment, 305:187–191, 1991.

Combination of various data analysis techniques for efficient track reconstruction in very high multiplicity events

Ferenc **Siklér**[1,a]

[1] *Wigner RCP, Budapest, Hungary*

Abstract. A novel combination of established data analysis techniques for reconstructing charged-particles in high energy collisions is proposed. It uses all information available in a collision event while keeping competing choices open as long as possible. Suitable track candidates are selected by transforming measured hits to a binned, three- or four-dimensional, track parameter space. It is accomplished by the use of templates taking advantage of the translational and rotational symmetries of the detectors. Track candidates and their corresponding hits, the nodes, form a usually highly connected network, a bipartite graph, where we allow for multiple hit to track assignments, edges. In order to get a manageable problem, the graph is cut into very many minigraphs by removing a few of its vulnerable components, edges and nodes. Finally the hits are distributed among the track candidates by exploring a deterministic decision tree. A depth-limited search is performed maximizing the number of hits on tracks, and minimizing the sum of track-fit χ^2. Simplified but realistic models of LHC silicon trackers including the relevant physics processes are used to test and study the performance (efficiency, purity, timing) of the proposed method in the case of single or many simultaneous proton-proton collisions (high pileup), and for single heavy-ion collisions at the highest available energies.

1 Introduction

Traditional methods of track reconstruction can be scaled to work in high multiplicity events, namely in many simultaneous collisions (pileup) of elementary particles [1, 2] and in high multiplicity single heavy-ion collisions. Nevertheless the performances are not optimal, efficiency and purity are reduced, especially at low momentum. That is why present data taking conditions and further luminosity and energy upgrades of high energy particle colliders, as well as those of detector systems, call for new ideas.

Image transformation methods and neural networks [3] are often used in gas detectors (time projection chambers [4, 5] and transition radiation trackers [6, 7]). In the case of silicon trackers the combinatorial track finding methods employed for trajectory building mostly use local information [8, 9]. They start with a trajectory seed and build a trajectory by extending the seed through the detector layers, picking up compatible hits. In the case of very many compatible hits the number of concurrently built trajectory candidates must be limited. Only some of the best candidates are kept which biases the final result. In this sense decisions are made too early. In addition, trajectories are mostly treated separately, there is no interaction between their assigned hits.

[a]e-mail: sikler.ferenc@wigner.mta.hu

In this study a combination of established data analysis techniques for charged-particle reconstruction is proposed. It uses all information available in an event while keeping competing choices open as long as possible.

2 Silicon detectors at particle colliders

At currently operating particle colliders the interaction region is very narrow (of the order of 50 μm) in both x and y (transverse) directions, while in z (longitudinal or beam) direction it is long, with a characteristic size of about 10 cm [10]. For single heavy-ion collisions the z_0 position of the primary interaction (vertex) is estimated with good precision using the copiously produced high p_T particles, thanks to the small pointing uncertainty of their tracks, reconstructed with traditional methods. In the case of single or multiple pp collisions no such information on the locations of the interaction vertices exists.

The trajectory of a primary particle is primarily determined by its initial position $(0, 0, z_0)$ and parameters (q, η, p_T, ϕ_0) of its initial momentum at creation. Here q is the charge, $\eta = -\ln\tan(\theta_0/2)$ is the pseudorapidity, p_T is the transverse momentum, ϕ_0 and θ_0 are the azimuthal and polar angle of the initial momentum vector in spherical coordinates. In a large volume solenoid the magnetic field near the center of the detector is rather homogeneous and points in the z direction. Hence in small volumes the trajectories of charged particles can be approximated by piecewise helices. For practical purposes a primary particle is parametrized by $(k_T, \sinh\eta, \phi_0, z_0)$ in the following, where $k_T = q/R$ is the signed curvature of the projection of its trajectory on the transverse (bending) plane, R is its radius. If the particle is singly charged the curvature is connected to p_T as $p_T = eB_zR$, where e is the electric charge of a proton.

Silicon trackers generally consist of several concentric cylindrical layers. Those close to the nominal interaction point are equipped with tiny pixel sensors, while others contain long strip sensors. Some strip layers are double-sided, they are located very close to each other two by two, and have a small relative tilt angle. The main characteristics of the inner barrel silicon detectors of the studied experimental setups are given in Table 1.

The trajectory of the primary particle intersects the concentric cylindrical layers and leaves hits behind in the silicon (Fig. 1). In the simplified case, when the magnetic field is homogeneous and if

Table 1. The main characteristics of the inner barrel silicon detectors of the studied experimental setups.

	B_z [T]	Layer type	Radii [cm]	Tilt [mrad]	$\sigma_{r\phi}$	σ_z (l_z)	x/X_0 [%]
Exp A	2.0	pixels	5.0, 8.8, 12.2	–	10 μm	115 μm	4
		strips	29.88, 29,92	±20	17 μm	(6.4 cm)	2
		strips	37.08, 37,12	±20	17 μm	(6.4 cm)	2
		strips	44.28, 44,32	±20	17 μm	(6.4 cm)	2
Exp B	0.4	pixels	3.9, 7.6	–	12 μm	100 μm	1
		drifts	14.9, 23.8	–	35 μm	25 μm	1
		strips	38.48, 38.52	+7.5, −27.5	20 μm	(4 cm)	0.5
		strips	43.58, 43.62	+7.5, −27.5	20 μm	(4 cm)	0.5
Exp C	3.8	pixels	4.4, 7.3, 10.2	–	15 μm	15 μm	3
		strips	25.48, 25,52	±50	23 μm	(10 cm)	2
		strips	33.88, 33.92	±50	23 μm	(10 cm)	2
		strips	41.8	0	35 μm	(10 cm)	2
		strips	49.8	0	35 μm	(10 cm)	2

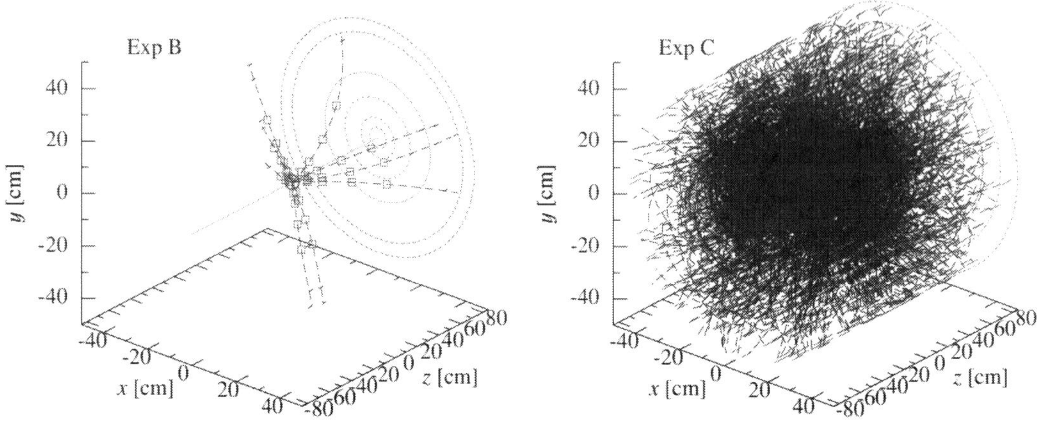

Figure 1. Left: Pixel hits (red open squares) and strip hits (red solid sections) of a single inelastic pp collision from simulation of Exp B. The location of the primary interaction is plotted with a green circle, the beam-line is indicated by a gray straight line. Charged particle trajectories (blue dashed curves) are also plotted. Right: Event with 40 simultaneous inelastic pp collisions from simulation of Exp C.

the detector material and its physical effects are neglected, the position of those hits could be precisely determined by simple equations. The physical effects of detector material changes this overly simple picture.

2.1 Physical effects

When a long lived charged particle propagates through material the most important effects which alter its momentum vector are multiple scattering and energy loss. The distribution of multiple Coulomb scattering is roughly Gaussian [11], the standard deviation of the planar scattering angle is

$$\theta_0 = \frac{13.6\ \text{MeV}}{\beta c p} z \sqrt{x/X_0}[1 + 0.038 \ln(x/X_0)], \tag{1}$$

where p, βc, and z are the momentum, velocity, and charge of the particle in electron charge units, and x/X_0 is the thickness of the scattering material in radiation lengths. Momentum and energy is lost during traversal of sensitive detector layers and support structures. To a good approximation the most probable energy loss Δ_p, and the full width of the energy loss distribution at half maximum Γ_Δ [12] are

$$\Delta_p = \xi \left[\ln \frac{2mc^2\beta^2\gamma^2\xi}{I^2} + 0.2000 - \beta^2 - \delta \right], \qquad \Gamma_\Delta = 4.018\xi, \tag{2}$$

where $\xi = \frac{K}{2} z^2 \frac{Z}{A} \rho \frac{x}{\beta^2}$ is the Landau parameter; $K = 4\pi N_A r_e^2 m_e c^2$; m is the mass of the particle; Z, A, I, and ρ are the mass number, atomic number, excitation energy, and the density of the material, respectively [11]. The density correction δ is neglected.

2.2 Hit clusters

An incoming charged particle loses energy in the sensitive detector elements by producing electron-hole pairs. The neighboring channels collecting a charge above a given threshold are grouped to form a cluster, a reconstructed hit. The size (dimensions) of the cluster depends on the angle of incidence of the particle: bigger angles will result in larger clusters. Due to the large fluctuations in energy loss, the measured $w_{r\phi}$ and w_z dimensions of the clusters will differ from the expected ones. In order to model these effects in the simulation, the cluster dimensions are varied by one unit in both directions for pixels, and up and down by two units in $r\phi$ direction for strips. In addition, if the size of a pixel cluster is at least two units in both directions, the layout of pixels with charge deposit and its location relative to the nominal interaction point usually indicate the sign of the electric charge of the particle. This way a pixel cluster is characterized by the measured widths $w_{r\phi}$ and w_z (dimensions of its rectangular envelope), and the charge, which can be ± 1 or left unknown. A strip cluster has only one such quantity, its $w_{r\phi}$ width.

2.3 Particle tracking

The Kalman filter is widely used in particle physics experiments for charged track and vertex finding and fitting and provides a coherent framework to handle known physical effects and measurement uncertainties [13]. It is equivalent to a global linear least-squares fit which takes into account all correlations coming from process noise. It is the optimum solution since it minimizes the mean square estimation error.

The state vector $x = (k, \theta, \psi, r\phi, z)$ is five dimensional, consisting of the signed inverse momentum, local polar angle, local azimuthal angle, global azimuthal position, and global longitudinal position, respectively. The propagation function $f(x)$ from layer to layer is calculated analytically using a helix model. Multiple scattering and energy loss in tracker layers is implemented with their Gaussian approximations shown in Eqs. (1)–(2). The propagation matrix $F = \partial f/\partial x$ is obtained by numerical derivation. The measurement vector for pixel hits $m = (r\phi, z)$ is two dimensional, for strip hits $m = (r\phi)$ it is one dimensional. The covariance of the process noise is $Q = (F_k \otimes F_k^T)\sigma_k^2 + (F_\theta \otimes F_\theta^T)\sigma_\theta^2 + (F_\psi \otimes F_\psi^T)\sigma_\psi^2$, where $\sigma_k = k\sigma_\Delta/\beta$, $\sigma_\theta = \sigma_\psi = \theta_0$ and $F_a = \partial f/\partial x_a$ is a vector. Multiple scattering contributes both to the variation of θ and ψ, while energy loss affects only k. The covariance of measurement noise V is a function of $\sigma_{r\phi}^2$ and σ_z^2, with slightly differing form for pixels and strips, where this latter also incorporates the value of the tilt angle.

Simulated particles are tracked while they are in the volume of the tracker detector, that is, trajectories looping in the magnetic field are properly followed. In the case of pattern recognition and track reconstruction only inside out propagation is considered.

Table 2. Ranges and the optimized number of bins (working point) corresponding to track parameters. The value of $p_{\mathrm{T,min}}$ is 0.1 GeV/c.

Variable	Range	Bins
k_T	$[-eB_z/p_{\mathrm{T,min}}, eB_z/p_{\mathrm{T,min}}]$	50
$\sinh\eta$	$[-\sinh(1.5), \sinh(1.5)]$	100
ϕ_0	$[-\pi, \pi]$	200
z_0	$[-3\sigma_z, 3\sigma_z]$	50

3 Pattern recognition

Our goal is to collect as much information as possible about potential track candidates, based on the location and shape of the measured hits in an event. To accomplish this, the position of each hit is transformed to a four-dimensional (k_T, sinh η, ϕ_0, z_0) accumulator space of track parameters. The accumulator space is not continuous but binned. (Bins are consecutive, adjacent, non-overlapping equal size intervals of a variable.) Ranges and the optimized number of bins corresponding to track parameters are shown in Table 2. The transformation is a variant of the well known Hough transform [14]. In the absence of physical effects (Sec. 2.1) the image of a point-like ($r\phi$, z) hit would be a well-defined two-dimensional manifold in that space, while the image of a section-shaped strip hit would be a three-dimensional manifold.

3.1 Preparation of templates

The detector models studied here have translational symmetry in longitudinal (z) and rotational symmetry in azimuthal ($r\phi$) direction. The $\phi - \phi_0$ angular difference primarily depends on k_T, while the $z - z_0$ longitudinal difference is mostly a function of sinh η (Fig. 2). These difference distributions further depend on the shape of the hit cluster (Sec. 2.2). For particles with a given (k_T, sinh η) and mass, the ($\phi - \phi_0$, $z - z_0$) values on a given detector layer populate a small rectangular area. The dimensions of that area result from the binning of the track parameters.

With help of numerous simulated particles we determine the populated area with help of local linear approximations around the center of each (k_T, sinh η) bin. In practice the central values and the Jacobian is deduced for each bin. The set of these values will be referred to as templates in the following. In order to have uniform coverage in all bins, the distribution of simulated particles is chosen

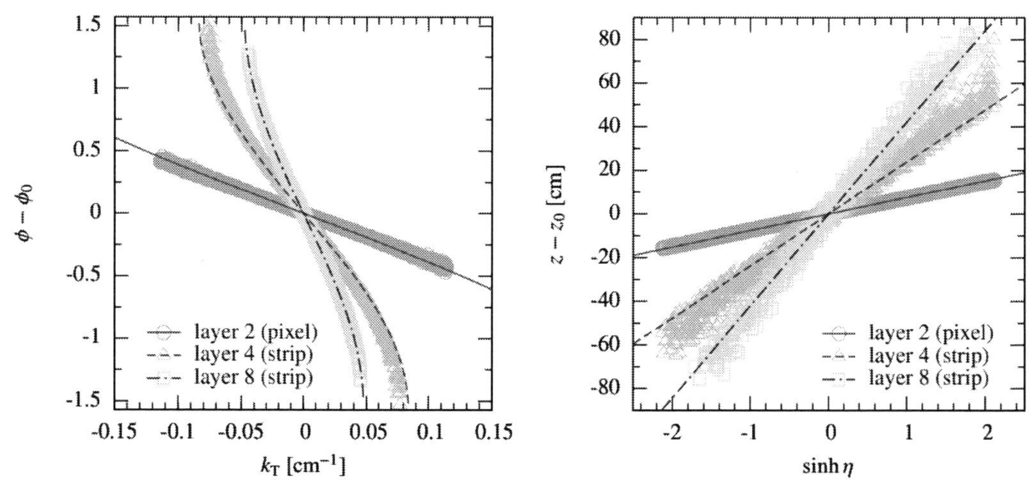

Figure 2. Left: Distributions of $\phi - \phi_0$ differences of the hit and track azimuth angles as a function of k_T for some selected detector layers. Right: Distributions of $z - z_0$ differences of the hit and track longitudinal coordinates as a function of sinh η for some selected detector layers. Values from simulation of Exp C are plotted with various symbols, while the oversimplified expectations $\phi - \phi_0 \approx -\arcsin(r/2 \cdot k_T)$ and $z - z_0 \approx 2 \sinh \eta \cdot \arcsin(r/2 \cdot k_T)/k_T$ at $k_T = 0.01$ cm^{-1} are shown with the curves.

to be constant in k_T, $\sinh\eta$, and ϕ_0. To limit fluctuations, normal distributed random variables, used in the simulation of physics processes (Sec. 2.1), are limited to values within 3.5 standard deviations (only about 0.05% lies outside this range). Altogether 2×10^6 pions are generated.

The prepared templates are used in two ways. During the early stage of image transformation they provide a (ϕ_0, z_0) accumulator area to increment for each (pixel) hit, in the case of a given $(k_T, \sinh\eta)$ bin. Later they are used to specify a search rectangle on the $(r\phi, z)$ plane of a (strip) layer for a given $(k_T, \sinh\eta, \phi_0, z_0)$ bin.

Although detector models with only barrel silicon detectors are studied here, the above considerations can be adopted to other geometries, such as disks perpendicular to the beam axis. In that case, the translational symmetry would be lost and the templates would become more complex by introducing another dimension, namely the relative z position of the primary interaction with respect to the longitudinal coordinate of the disk.

3.2 Image transformation

The transformation of spatial information to track space proceeds as described in the following. First hits on the three innermost layers are dealt with, containing exclusively pixel hits. For each a hit all potential $(k_T, \sinh\eta)$ accumulator bins are examined and the corresponding possible (ϕ_0, z_0) values, enveloped by rectangles, are determined. Bins within such $(k_T, \sinh\eta, \phi_0, z_0)$ area are incremented. Since we look for tracks with hits on all three innermost layers, only those accumulator bins are kept which gathered votes from all three layers.

The subsequent layers usually contain strip hits. The method used for the innermost layers would not be efficient here because there are far too many accumulator bins to handle. To this end, the kept accumulator bins are examined, corresponding to proto-tracks with three counts, obtained in the previous step. Using the bin coordinates $(k_T, \sinh\eta, \phi_0, z_0)$ we look for compatible hits by determining a search rectangle on the $(r\phi, z)$ plane for each layer. For quick access, and in order to facilitate hit selection, strip hits are in advance partitioned on an equidistant grid using their $(r\phi, z)$ coordinates.

The search for compatible hits proceeds outwards. It is advantageous since the process can be abandoned if some layers provided no compatible hits while they are still reachable according to the curvature range of the examined bin. In other words, the number of layers with compatible hits should not be very different from the number of reachable layers. (We can allow for a few layers without hits.)

3.3 Trajectory building

Track candidates are built using hits collected in given accumulator bins. Trajectory propagation and track fitting is performed by the extended classical Kalman filter [13] including prediction, filtering, and smoothing, with pion mass assumption. The initial state vector is estimated by placing a helix to the innermost two hits and using the beam-line as a constraint by adding a zeroth point with value $r\phi = 0$, and with an uncertainty of $\sigma_{r\phi} \approx 50\ \mu m$. In the case of an off-centered beam, $\sigma_{r\phi}$ can be increased to properly contain the interaction region in the transverse plane.

Trajectory building normally proceeds from inside out and considers all hit combinations recursively by forming branches. At each layer there are usually multiple hits to add to the existing trajectory. The number of hits to be considered is especially large for the inner strip layers that would exponentially increase the number of trajectory branches.

A powerful solution for this problem is the effective detection of unsuitable (outlier) hits in the busy detector layers. For this purpose, trajectory building starts with the estimate of the initial state vector at the zeroth point, which is then propagated through the first three layers. At that point we

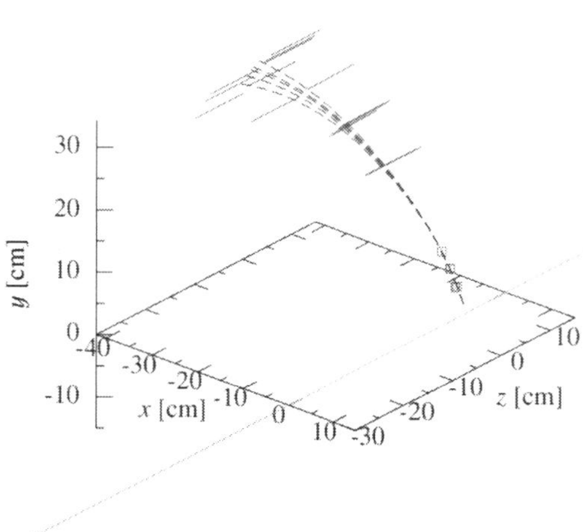

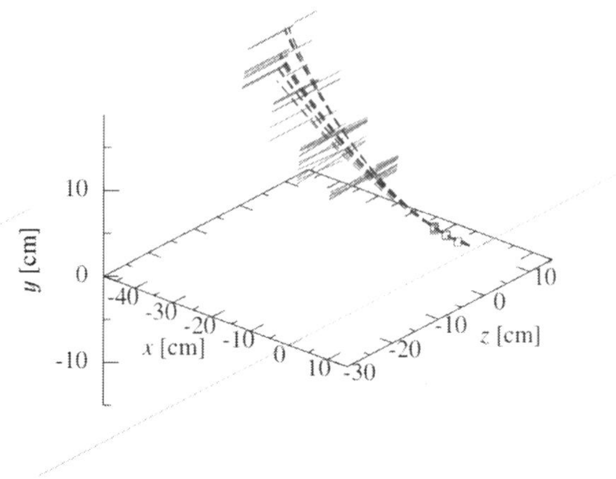

Figure 3. Hits (red boxes and line sections) belonging to a given bin in the accumulator space and trajectories propagated to the outermost detector layer (blue dashed curves). The beamline is indicated with the gray line.

already have a good knowledge about the parameters of the track that is being reconstructed. Instead of going to the next (strip) layer, the trajectory is at once propagated to one of the potential hits in the outermost detector layer (Fig. 3). During propagation the physical effects of the crossed layers are duly taken into account, but information about their hits is not used. The outliers in the intermittent omitted layers are detected next in the smoothing step of the deficient trajectories using the smoothed residual [13]. Those (strip) hits in the upper 0.5% tail of the corresponding χ^2 distribution are discarded.

Once the list of compatible strip hits is narrowed down, full trajectory building with the selected hits is performed again. During trajectory building we can allow for a few (one or two) missing hits. It may mean no hit at all or the case when χ^2 is too large for a given number of degrees of freedom (ndf). If there are too many missing hits, the process is abandoned. In order not to lose a noticeable amount of track candidates, but also to have a good selection power, trajectories in the upper 0.5% tail of the corresponding χ^2 distribution are discarded and not developed further.

4 Optimal distribution of hits among tracks

In the end we have a set of track candidates with the somewhat unusual property that temporarily several track candidates may share some hits. Our goal is to resolve these ambiguities, hit confusion, by optimally allocating the hits among tracks, since all hits must be assigned to not more than one track. The hits and track candidates, and their relations, are best represented by a bipartite graph G. The nodes of G are from two disjoint sets: hits and track candidates, such that each edge connects a hit to a track candidate (Fig. 4). Our goal is to assign all hits to at most one track, while keeping an eye on the total merit of all tracks ($\sum \chi^2$) in the event.

First those track candidates are privileged which are very likely real. To this end, we extract a subgraph selecting track candidates which have at least three hits not requested by other candidates, that is, at least three leaf hit nodes. (A leaf hit node is connected to exactly one track candidate node.) This subgraph is disconnected (Sec. 4.1), the arising minigraphs are solved (Sec. 4.2) individually. Selected tracks and stored, edges and nodes of the subgraph are removed from G. Next we extract the subgraph containing track candidates with the highest maximum number of possible hits (usually

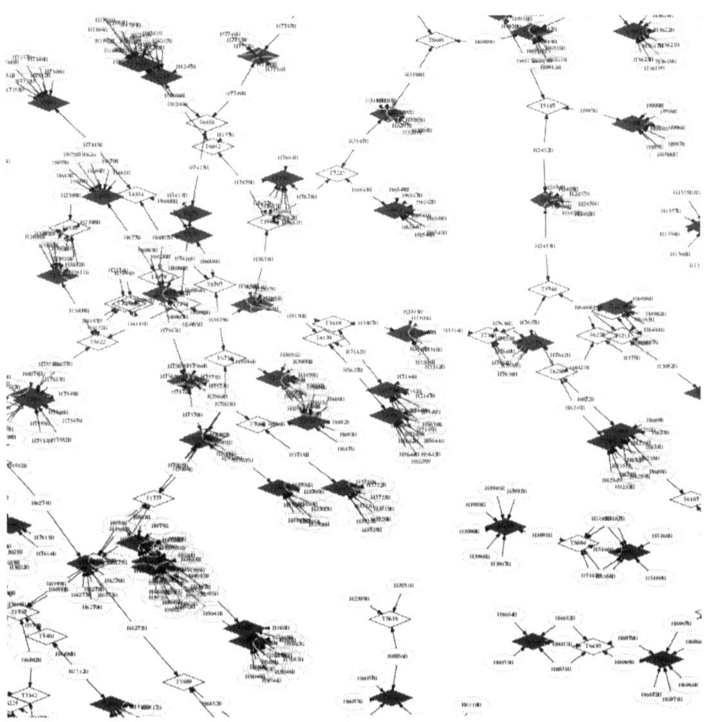

Figure 4. A small fraction of the bipartite graph G of hits (ellipses) and track candidates (blue diamonds) for a multiple pp collision event (40 simultaneous inelastic pp collisions). Directed arrows, graph edges, show potential hits-to-track candidate assignments.

$n = 9$), and their corresponding hits. Thanks to the high number of hits required, these track candidates are likely real. Selected tracks are again stored, nodes and edges removed from G as above. Then the process is restarted with the subgraph of track candidates with $n - 1$ hits, iterating down to the subgraph of track candidates with three hits.

4.1 Disconnecting a subgraph

The graphs encountered are usually highly connected. If we would try to allocate the hits to tracks one by one, the number of trials needed would explode exponentially with increasing number of nodes. In order to reduce the complexity of the problem the graphs should be partitioned into several small pieces. This can be accomplished by finding vulnerable components in the graph whose removal disconnects the graph. Such vulnerable elements are special edges (bridges) and special nodes (articulation points) whose deletion increases the number of connected components of the graph. Of course this way some tracks would lose a hit, but that is only a small price to pay.

Bridges and articulation points can be found in linear time with help of graph traversal techniques. The depth-first search is an algorithm for traversing and searching a graph. One starts at some arbitrary node as the root and explores as far as possible along each branch before backtracking. The nodes of bridges and the articulation points are found by requiring that their children nodes do not have a backedge. In such a "disconnecting" step bridges and articulation points are removed. During the

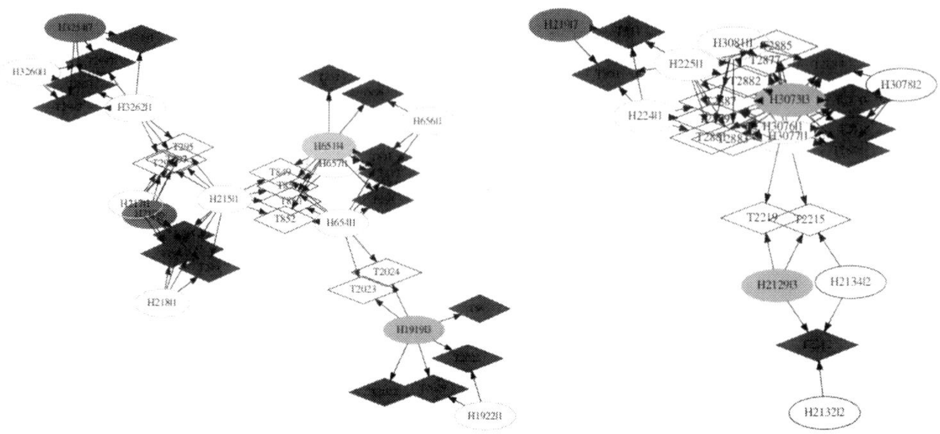

Figure 5. Example minigraphs obtained after the removal of all bridges and articulation points from the bipartite graph G of hits and track candidates, in the case of a multiple pp collision event (40 simultaneous inelastic pp collisions). The colors of contracted hits (ellipses) refer the number of hits with identical role (red – 6 or more, orange – 4 or 5, green – 3, white – 1 or 2). Filled track candidates (blue diamonds) indicate true tracks, while the others (open diamonds) show candidates where one or more hits are not in place.

process new vulnerable elements may come to light, hence the disconnecting steps are repeated until no new such elements are found.

In addition, the resulted graph is further partitioned into disjoint graphs. This task is best accomplished by the flood fill method embedded into the above detailed traversal technique. The output of the disconnecting step is a large set of disjoint minigraphs.

In high-pileup pp collision events there are usually several thousand track candidates. Their corresponding bipartite graph G and its subgraphs contain several hundred bridges and up to 50 articulation points. Once the subgraphs are disconnected, we get couple of thousand minigraphs.

4.2 Solving a minigraph

A minigraph usually has several hits (between 2 and 7 hits for the detector models studied here) with identical role: they are connected to the same set of track candidates. In the interest of reducing complexity hits with identical role are treated jointly, the set of such hits is "contracted" (Fig. 5).

The number of remaining nodes is usually small, the contracted hits can be distributed among tracks by building and solving a decision tree. The process is similar to exploring decision trees of deterministic, strategy board games, such as chess and go, including their horizon problem (limited search depth). What is different here is that our process is a single-player one. The optimal hit-to-tracks assignments are chosen in the following recursive method:

1. First the most important, highest ranked available (contracted) hit is located. Rank is calculated as the product of the number of similar hits (the number of hits the contracted hit represents) and the number of edges the hit node has (the number of associated track candidates). Such a definition gives preference to contracted hits which represent many particle hits and have a central role in the graph.

DOI: 10.1051/epjconf/201715000011

2. The highest ranked hit can be attached to several track candidates, and these choices are evaluated sequentially and recursively as branches of a decision tree. After a hit-track assignment is chosen, the track and its hits are selected, and their nodes and all corresponding edges are removed from the minigraph.

3. Track candidate nodes and corresponding edges with too few remaining hits (less than three) and those with too many missing hits are also removed.

4. As long as there are nodes left in the minigraph we go back to step 1. Otherwise the actual path of the decision tree is evaluated based primarily on the amount of hits on selected tracks. If there are two decision trees with the same amount of hits, the one with lower $\sum \chi^2$ of the selected tracks is chosen.

In order to save time during solving the decision tree, the selected tracks candidates are not refitted but only the adjusted χ^2 values of their remaining hits, calculated based on a Kalman-fit using all the initial hits, are summed and the corresponding ndf values are recalculated.

In the end the decision path with the best score is taken, the selected tracks and their hits are stored.

5 Results

In the computer simulation the interaction region is centered at $(x, y, z) = (0, 0, 0)$. In z (beam) direction it is described by a Gaussian distribution with a standard deviation of $\sigma_z = 5$ cm. The silicon tracker covers the pseudorapidity range of $|\eta| < 1.5$. Inelastic pp collisions at $\sqrt{s} = 14$ TeV are obtained from the PYTHIA8 [15] Monte Carlo event generator (version 219). It is known to well reproduce measured momentum spectra of charged particles in pp collisions at $\sqrt{s} = 13$ TeV [16–18] with average pseudorapidity density of $dN/d\eta \approx 5.5$ near $\eta \approx 0$, as well as the composition of the most abundant charged particles (pions, kaons, protons). Semi-central PbPb collisions at $\sqrt{s_{\mathrm{NN}}} = 5.5$ TeV with $dN/d\eta \approx 1200$ are obtained from the HYDJET [19] Monte Carlo events generator (version 1.9). It was retuned to provide measured momentum spectra of charged particles in the highest energy heavy ion collisions, as seen in $\sqrt{s_{\mathrm{NN}}} = 5.02$ TeV energy central PbPb collisions [20]. Physical effects (multiple scattering and energy loss) are simulated according to the simple models detailed in Sec. 2.1 using the description of detector materials shown in Table 1.

The proposed algorithm is coded in C++ and run on a 2.1 GHz dual-core machine. The average CPU time needed was on average 25 sec for events with 40 simultaneous inelastic pp collisions. Its performance as a function of transverse momentum p_{T} of the charged particles is shown in Fig. 6. (A reconstructed track is considered matched to a simulated one if all their hits correspond to each other, or if at most one of them is not in place.) Efficiency and fake track rate are plotted for two experimental setups. The values are separately given for single pp, for multiple (40) simultaneous pp, and for semi-central PbPb collisions. It is clear that for $p_{\mathrm{T}} > 0.2$–0.4 GeV/c the efficiency is above 90–95% and fake track rate is well below 1%, independent of collision system (pp, PbPb) and pileup (1–40). At very low transverse momentum ($p_{\mathrm{T}} < 0.1$–0.2 GeV/c) efficiency drops and fake track rate increases to some 2–4%. Both measures show a clear advantage and convincing performance of the proposed methods over those presently used in the highest energy particle physics experiments.

In the case of multiple pp collisions most computing time is spent on the image transformation, while for PbPb trajectory building takes most of the resources. Disconnecting and solving the graph is quick in both cases. Efficiency, running times, and fake track rate values as a function of the number of simultaneous inelastic pp collisions are shown in Fig. 7. Efficiency for all charged particles slowly

DOI: 10.1051/epjconf/201715000011

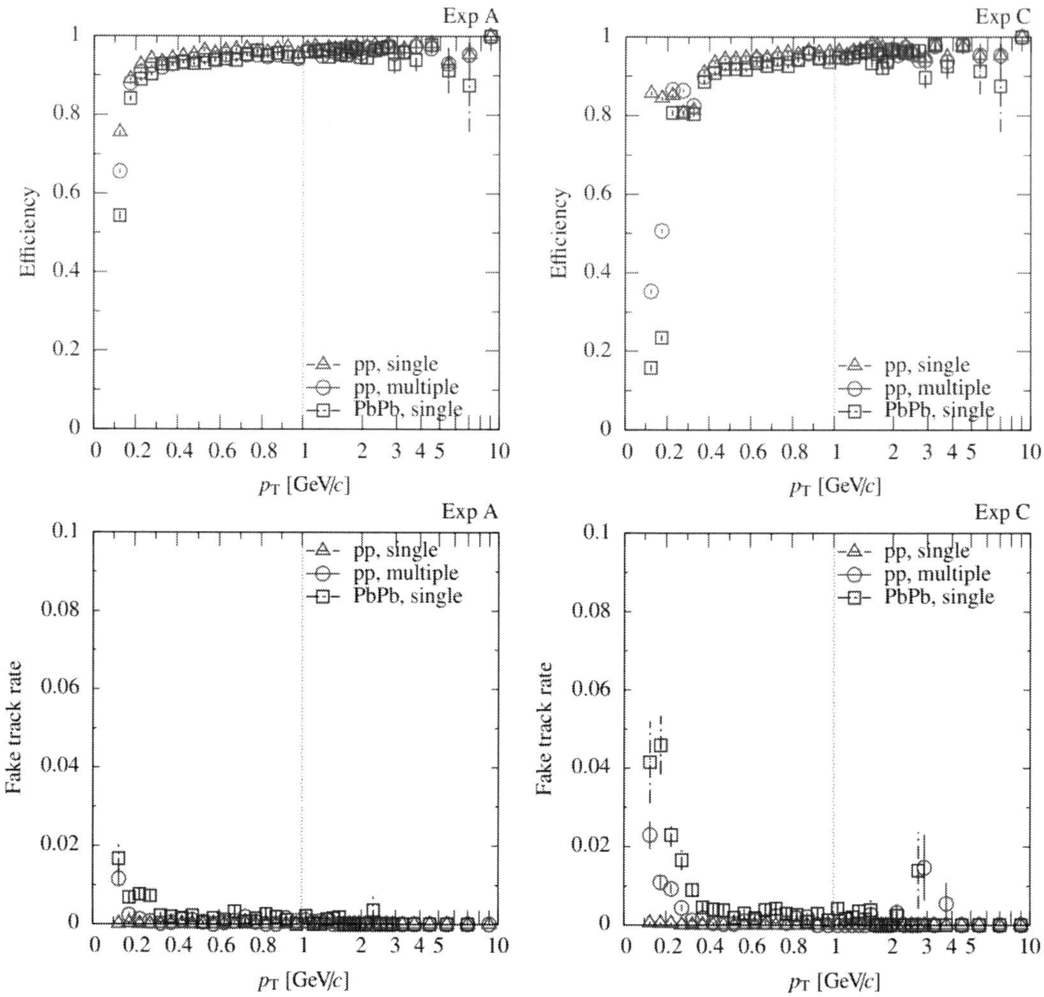

Figure 6. Performance of the proposed algorithm as a function of transverse momentum p_T of the changed particles. Efficiency (top) and fake track rate (bottom) are plotted for Exp A (left) and Exp C (right). The values are separately given for single pp (open green triangles), multiple (40) pp (open red circles), and semi-central PbPb (open blue boxes) collisions. The horizontal scale is linear in the region 0–1 GeV/*c*, while it is logarithmic for 1–10 GeV/*c*.

decreases but stays in the 80–90% range. The fake track rate starts at the permille level and stays under a percent.

6 Summary

A combination of established data analysis techniques for charged-particle reconstruction was presented. It follows a global approach and uses all information available in a collision event. It employs image transformation based on precomputed templates taking advantage of the translational and ro-

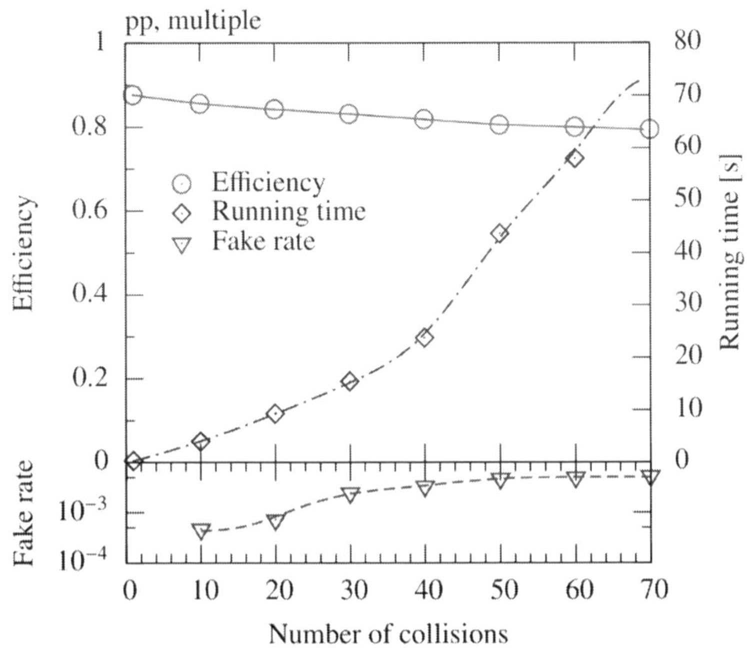

Figure 7. Performance of the proposed algorithm as a function of the number of simultaneous inelastic pp collisions. Efficiency (open red circles, top left), running times (open blue diamonds, top right), and fake track rate (open green triangles, bottom) values are plotted. Lines are drawn to guide the eye.

tational symmetries of the detectors. Track candidates and their corresponding hits usually form a highly connected network, a graph. The graph is partitioned into very many minigraphs by removing a few of its vulnerable components, edges and nodes. The hits of a subgraph are distributed among the track candidates by solving a deterministic decision tree.

Tests using simplified computer models of LHC silicon trackers show that efficiency and purity of track reconstruction are excellent and the timing of the proposed method is reasonable, both in simultaneous proton-proton collisions (high pileup), and in single heavy-ion collisions at the highest available energies.

Acknowledgements

The author wishes to thank to Sándor Hegyi and András László for helpful discussions. This work was supported by the Swiss National Science Foundation (SCOPES 152601), and the National Research, Development and Innovation Office of Hungary (K 109703).

References

[1] G. Aad et al. (ATLAS), Eur. Phys. J. C **76**, 581 (2016), 1510.03823
[2] M. Rovere (CMS), J. Phys. Conf. Ser. **664**, 072040 (2015)
[3] R. Fruhwirth, Comput. Phys. Commun. **78**, 23 (1993)

[4] K. Aamodt et al. (ALICE), JINST **3**, S08002 (2008)
[5] C. Cheshkov, Nucl. Instrum. Meth. A **566**, 35 (2006)
[6] G. Aad et al. (ATLAS), JINST **3**, S08003 (2008)
[7] B. Mindur (ATLAS), Nucl. Instrum. Meth. A **845**, 257 (2017)
[8] A. Strandlie, Nucl. Instrum. Meth. A **535**, 57 (2004)
[9] S. Chatrchyan et al. (CMS), JINST **3**, S08004 (2008)
[10] M. Lamont, Journal of Physics: Conference Series **455**, 012001 (2013)
[11] C. Patrignani et al. (Particle Data Group), Chin. Phys. C **40**, 100001 (2016)
[12] H. Bichsel, Rev. Mod. Phys. **60**, 663 (1988)
[13] R. Fruhwirth, Nucl. Instrum. Meth. A **262**, 444 (1987)
[14] P.V.C. Hough, Tech. rep. (1962), US Patent 3069654
[15] T. Sjöstrand, S. Mrenna, P.Z. Skands, Comput. Phys. Commun. **178**, 852 (2008), 0710.3820
[16] V. Khachatryan et al. (CMS), Phys. Lett. B **751**, 143 (2015), 1507.05915
[17] J. Adam et al. (ALICE), Phys. Lett. B **753**, 319 (2016), 1509.08734
[18] G. Aad et al. (ATLAS), Phys. Lett. B **758**, 67 (2016), 1602.01633
[19] I.P. Lokhtin, A.M. Snigirev, Eur. Phys. J. C **45**, 211 (2006), hep-ph/0506189
[20] J. Adam et al. (ALICE), Phys. Rev. Lett. **116**, 222302 (2016), 1512.06104

Parameterization-based tracking for the P2 experiment

Iurii **Sorokin**[1,2,a] for the P2 collaboration

[1] *Institute for Nuclear Physics, University of Mainz, Germany*
[2] *PRISMA Cluster of Excellence*

Abstract. The P2 experiment in Mainz aims to determine the weak mixing angle θ_W at low momentum transfer by measuring the parity-violating asymmetry of elastic electron-proton scattering. In order to achieve the intended precision of $\Delta(\sin^2\theta_W)/\sin^2\theta_W = 0.13\%$ within the planned 10 000 hours of running the experiment has to operate at the rate of 10^{11} detected electrons per second. Although it is not required to measure the kinematic parameters of each individual electron, every attempt is made to achieve the highest possible throughput in the track reconstruction chain.

In the present work a parameterization-based track reconstruction method is described. It is a variation of track following, where the results of the computation-heavy steps, namely the propagation of a track to the further detector plane, and the fitting, are pre-calculated, and expressed in terms of parametric analytic functions. This makes the algorithm extremely fast, and well-suited for an implementation on an FPGA.

The method also takes implicitly into account the actual phase space distribution of the tracks already at the stage of candidate construction. Compared to a simple algorithm, that does not use such information, this allows reducing the combinatorial background by many orders of magnitude, down to $O(1)$ background candidate per one signal track.

The method is developed specifically for the P2 experiment in Mainz, and the presented implementation is tightly coupled to the experimental conditions.

1 The P2 experiment

1.1 Motivation

One of the fundamental parameters in the Standard Model is the weak mixing angle θ_W, which defines the relative strength of the electromagnetic and weak interactions. The weak mixing angle is a scale-dependent quantity and it has been measured at various energies and in different processes [1] (Fig. 1). However, the most precise measurements, performed at the Z pole, are marginally consistent, and the measurements at lower energies are not precise enough to discriminate between the Standard Model and possible extensions. The P2 experiment in Mainz aims at measuring the weak mixing angle at low momentum transfer (average momentum transfer squared $Q^2 = 4.5 \cdot 10^{-3}$ GeV^2/c^2) to a precision of 0.13%. Such measurement will be a stringent test of the Standard Model, and has the potential of discovering physics beyond the Standard Model in the running.

[a] e-mail: sorokin@uni-mainz.de

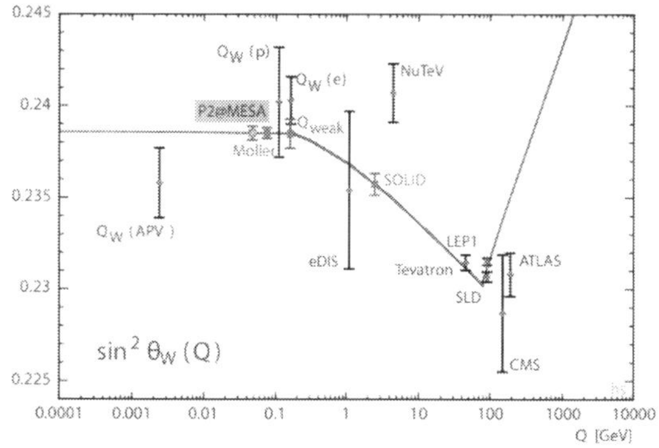

Figure 1. Scale dependence of $\sin^2\theta_W$ together with completed (black error bars) and planned (blue error bars, value chosen to coincide with theory) experimental measurements [1].

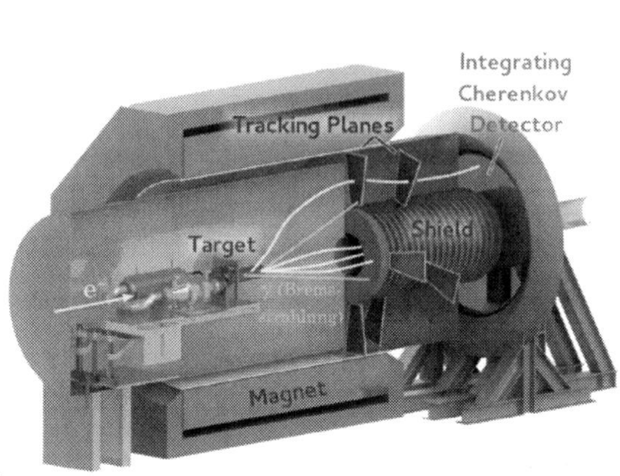

Figure 2. CAD model of the P2 detector.

1.2 The method and the setup

In the P2 experiment the weak mixing angle is going to be determined from the parity-violating asymmetry of elastic electron-proton scattering. The latter can be related to θ_W as follows:

$$A_{PV} = \frac{\sigma_L - \sigma_R}{\sigma_L + \sigma_R} = \frac{G_F Q^2}{4\sqrt{2}\pi\alpha}(Q_W + F(Q^2)), \quad (1)$$

$$Q_W = 1 - 4\sin^2\theta_W, \quad (2)$$

where A_{PV} is the parity violating asymmetry, σ_L and σ_R are the elastic scattering cross-sections for left- and right-handed electrons, Q_W is the weak charge of the proton, $F(Q^2)$ is the Fermi constant, α is the fine structure constant, and Q^2 is the momentum transfer squared.

The planned experimental setup is shown in Fig. 2. A longitudinally polarized electron beam is scattered in a liquid hydrogen target, and the current of the elastically scattered electrons is measured with a Cherenkov Detector. The beam polarization is flipped with a rate of several kHz. Providing the integrated luminosity with both helicities is the same, the parity violating asymmetry can be evaluated as:

$$A_{PV} = \frac{N_L - N_R}{N_L + N_R}, \quad (3)$$

where the N_L and N_R are the number of detected electrons with left-handed and right-handed helicities respectively.

As A_{PV} is on the order of 10^{-8}, to achieve the required statistical precision of $\Delta A_{PV} \leq 6 \cdot 10^{-10}$, the number of detected electrons needs to be higher than $3 \cdot 10^{18}$. In order to collect such immense statistics on a feasible timescale, assuming 10 000 running hours, the experiment needs to work at a rate of 10^{11} detected electrons per second.

A new accelerator, MESA [2], is going to be built to provide the required high intensity (150 μA), highly polarized (> 85%) beam with an exceptional stability: helicity-correlated position fluctuations below 0.13 nm, angular fluctuations below 0.06 nrad, and intensity fluctuations below 0.36 ppb. The energy of the beam is chosen to be 155 MeV to minimize the theoretical uncertainties [3].

The long (60 cm) target is also necessary to achieve the required event rate. A vast amount of bremsstrahlung photons is produced in the target, creating a background in the Cherenkov Detector. The lead shield in the middle of the setup blocks the photons coming directly from the target, but still a considerable amount of the photons hit the Cherenkov Detector after a Compton scattering in the inner surface of the magnet. The superconducting magnet bends the electrons around the shield.

Evaluating the weak mixing angle from equation 1 requires knowing the distribution of the momentum transfer squared Q^2. Because of the considerable energy loss and multiple scattering in the target, the Q^2 distribution need to be determined from a detailed simulation. Here one can not fully rely on GEANT4, as the precise shape of the scattering distribution is important, and helicity correlated effects are not implemented. To properly tune and validate the simulation the Tracking System is necessary.

1.3 Tracking system

The main purpose of the Tracking System is to help evaluating the Q^2 distribution of the primary electron-proton scattering. In addition, the Tracking System is going to be used for detector alignment, as well as for monitoring the experimental conditions by means of watching the kinematic distributions.

The Q^2 distribution needs to be determined with the relative systematic uncertainty on the mean below 0.2%. The measurement will be performed at a decreased beam rate, to reduce the effect of combinatorial background. With the same data sets the detector alignment and validation of the magnetic field map will be performed.

During the normal experiment operation, at the full beam rate, the amount of data, produced by the Tracking System will be too large to be processed with the affordable computing resources. The data will be acquired in short intervals, with sufficient pauses for processing in between. Every attempt is made to achieve the highest possible throughput in the reconstruction chain, as this would result in better sensitivity to variations of the experimental conditions. The minimum required duty cycle can not be quoted at the moment, as it depends on the tracking performance and details of experimental conditions.

The Tracking System will be based on the MuPix [4] high-voltage monolithic active pixel sensors (HV-MAPS). MuPix features high rate capability (2 MHz/chip tested, 284 MHz/cm^2 theoretical), time resolution of about 11 ns, low mass, low power consumption (300 mW/cm^2), and high efficiency (99.5%) [4]. The chip was originally developed for the Mu3e experiment [7], and fits very well to the needs of P2.

The sensors will be arranged in four planes. The choise of four planes was made to minimize the material budget. At least three planes are necessary to constrain the track parameters within a magnetic field, and at least one more is needed to distinguish between the real tracks and random hit combinations. The planes are placed in pairs, with 2 cm distance between the planes within each pair, and 54 cm between the pairs (Fig. 3). Placing the planes close to each other reduces the area to search for the matching hit, hence reduces the combinatorial background. Yet, a certain spacing is necessary to accurately measure the track curvature.

The Tracking System consists of four sectors (Fig. 2), each covering about 15° of acceptance in azimuth angle ϕ. Locating the sectors at 90° and 180° w.r.t. to each other is favorable to detect spatial asymmetries.

1.4 Reconstruction challenge

The major challenges in track reconstruction are the extreme track rate, high hit density, and considerable background from the bremsstrahlung photons. For the first estimates it was assumed that the track reconstruction is done in 45 ns time frames, which corresponds approximately to four times the time resolution of the MUPIX chip (measured to be σ_t=11 ns[4]). At the full beam rate this leads to about 800 reconstructible signal electrons and up to 6000 background hits per plane in one frame.

Obviously, at such high rates, the reconstruction has to be done on-line. To achieve the highest possible throughput the reconstruction is considered to be implemented on a FPGA. For this a fast and simple algorithm is required, such as the parameterization-based tracking.

1.5 Simulations

In the following sections references to the GEANT4 [5] simulations will be made. At present the GEANT4 model of the P2 experiment includes most of the relevant material in the detector, uses a realistic magnetic field map, and considers the relevant electromagnetic processes. The major simplifications in the model are that: (a) the incident beam is stationary, and goes along the z-axis, whereas in fact it is planned to scan with the beam over the target volume in order to prevent the target from boiling, (b) that the tracking planes are flat, without any dead area, whereas in fact they will consist of 2x2 cm^2 sensors, overlapping in x and y (because of the dead edges), and interleaved in z, and (c) the misalignment and the uncertainty on the field map are not simulated. The present model is adequate for evaluating the performance of the tracking algorithm. Further extension of the model are necessary for other studies, as will become clear from the text.

2 Parameterization-based tracking

2.1 Concept

The presented *parameterization-based tracking* is a variation of *track following*. Track following is a track reconstruction method, when first track seeds are constructed, and then hits from the consequent planes are added to the seeds one-by-one. Finally, the constructed hit combinations — *track candidates* — are fitted, and the best candidates are selected based on the fit quality information, usually χ^2.

In P2 the reconstruction goes backwards: each hit in plane 3 (Fig. 3) is taken as a track seed, and then hits from the planes 2, 1 and 0 are added in the respective order. A combination of one, two, or three hits, constructed in such way, will be called a *track segment*.

To decide which hits can be added to a track segment one needs to extrapolate the segment to the next (upstream) plane. The extrapolated position, along with the extrapolation uncertainty, define the *region of interest* (ROI) — the region where all hits match to the track segment. Usually, there is more than one hit in the ROI. In such case extended segments are constructed with each of the matching hits.

Extrapolation of a segment is associated with certain difficulties. An explicit extrapolation requires an estimate for the kinematic parameters of the track segment. The kinematic parameters, in particular the momentum, can be estimated only if the segment consists of three or more hits. To find matching hits for one- and two-hit segments one would need to use a predefined, sufficiently large ROI, which is inefficient in terms of combinatorial background. In addition, the explicit extrapolation requires looking up the geometry and the magnetic field map, as well as performing numerical integration, which all together can take a significant fraction of the overall computation time.

In the parameterization-based tracking the ROIs are pre-calculated beforehand, and expressed in terms of parametric functions of the coordinates of the segment hits. So, during the reconstruction, the optimal ROI can be easily evaluated from the parameterizations for any track segment.

In the current implementation, for the sake of simplicity, the ROIs are chosen to be rectangular. The characteristics of the rectangular ROI, namely the x- and y-positions, the x- and y-sizes, and the rotation are expressed as 3rd order polynomial functions of the segment hit coordinates. In the following these quantities will be referred to as x_{pos}, y_{pos}, x_{size}, y_{size}, and θ_{rot} respectively. How exactly x_{pos}, y_{pos}, x_{size}, y_{size}, and θ_{rot} depend on the coordinates of the segment hits will be explained in section 2.3. The variables x_{pos}, y_{pos}, x_{size}, y_{size}, and θ_{rot} express the position of the ROI respectively to the last hit in the segment, in a rotated reference frame, as shown in Fig. 4. The 3rd order polynomials were chosen also for simplicity, and planned to be replaced with splines in the future. Also, for technical reasons, in the current implementation θ_{rot} is always zero, which also needs to be fixed.

The next section (2.2) will describe how to define the ROI for a given track segment. In section 2.3 it will be shown how to proceed from a set of segments with defined ROIs to the sought parameterizations. In sections 2.4 and 3.3 it will be demonstrated that not only the ROIs, but also the track parameters, such as the position and the momentum, can be parameterized.

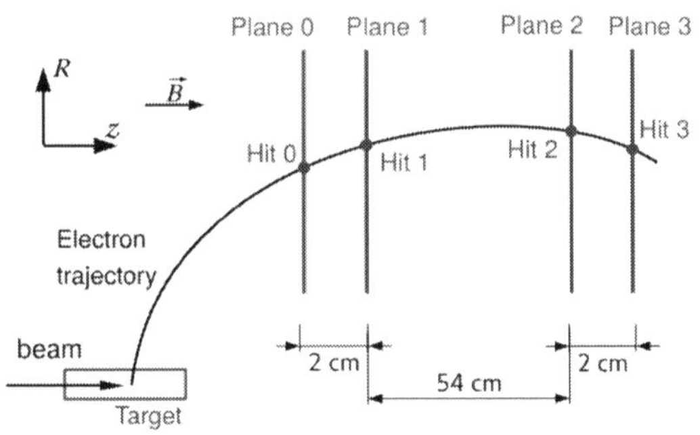

Figure 3. Schematic zR-view on the target and the tracking planes.

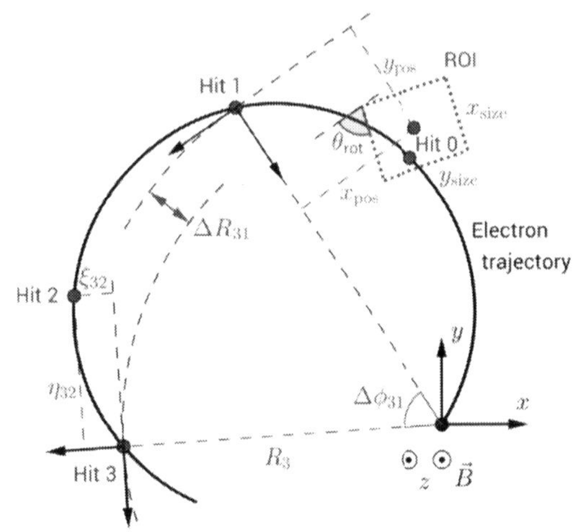

Figure 4. Schematic xy-view on the target and the tracking planes.

2.2 Defining the region of interest

In this work the ROIs are defined by comparing the segments with reference tracks. The reference tracks are either tracks from the GEANT4 simulation, or high quality (low χ^2) tracks, reconstructed from the low beam rate experimental data. At low beam rate the combinatorial background is small, so track candidates can be constructed with a simple exhaustive search, and a conventional rigorous fit of every candidate can be performed.

To define the ROI one first selects reference tracks that are similar to the given track segment. More precise, it is required that each hit of the reference track would lie within a certain range around the corresponding, i.e. lying in the same plane, hit of the segment. If the segment has no hit in a certain plane, then the corresponding hit of the reference track is ignored during the selection. Having the reference tracks selected in this way, it is trivial to define the ROI: it is the region, where the selected reference tracks are concentrated. In the current implementation it is defined as the rectangle of the minimal area, which encloses not less than a certain fraction (typically 95%–99%) of the reference tracks. Obviously, the ROI, defined in such way, will enable to find only these tracks, for which similar reference tracks were available. To go beyond this limitation one can scale up the size of the obtained ROI.

2.3 Parameterizing the region of interest

First the general form of the parameterizations need to be defined. As mentioned in section 2.1, in the current implementation the ROIs are chosen to be rectangular, so one needs to define five functions: x_{pos}, y_{pos}, x_{size}, y_{size}, and θ_{rot}.

The functions were chosen to be 3rd order polynomials, and must depend on the coordinates of the segment hits. But using the coordinates of all hits explicitly is disadvantageous because this would make the parameterization functions to be up to six-dimensional, which would later on lead to certain technical difficulties. The number of variables can be reduced by taking one symmetry and one correlation into account.

First the functions for the ROI in plane 2 (Fig. 3) will be defined. The ROI in plane 2 can depend only on the coordinates of the hit in plane 3. Since the position of the ROI is defined relativeley to the last hit of the segment (hit 3), and due to the rotational symmetry of the setup, the functions x_{pos}, y_{pos}, x_{size}, y_{size}, and θ_{rot} can be defined as functions of the radial position of hit 3, R_3, only (Fig. 4).

The ROI in plane 1 can depend only on hits in planes 3 and 2. Again, due to the rotational symmetry, the functions x_{pos}, y_{pos}, x_{size}, y_{size}, and θ_{rot} can be defined to depend only on R_3 and on the relative position of hits 2 and 3, ξ_{32} and η_{32} (Fig. 4).

The ROI in plane 3 should, in general, depend on position of hits in planes 3, 2, and 1. But the positions of hits 3 and 2 become strongly correlated as soon as the position of hit 1 is fixed. Indeed, the x- and y-position of hits 1 and 3, together with the target, make a coarse constraint on the track radius, whereas the z positions of the two hits, together with the defined track radius, constrain the track inclination w.r.t. to the xy-plane. Even the coarse constraints on the track parameters lead to a strong correlation between hits 3 and 2 because the planes are very close to each other. Therefore, in the current implementation the position of hit 2 is ignored. Taking into account the rotational symmetry, the functions x_{pos}, y_{pos}, x_{size}, y_{size}, θ_{rot} are defined to depend on R_3, as well as on the relative position of hits 1 and 3, ΔR_{31} and $\Delta\phi_{31}$ (Fig. 4). Here polar coordinates are chosen because they are less correlated.

Next step is to find the coefficients of the functions x_{pos}, y_{pos}, x_{size}, y_{size}, θ_{rot}. This is done by fitting the functions to ROIs, created from the reference tracks as follows. The phase spaces of the parameterization functions, $\{R_3\}$ in case of plane 2, $\{R_3, \xi_{32}, \eta_{32}\}$ in case of plane 1, and $\{R_3, \Delta R_{31}, \Delta\phi_{31}\}$ in case of plane 0, are divided in bins. All reference tracks are grouped according to each of the three binnings. For each bin in each phase space a ROI is created using the reference tracks in this bin (an example is shown in Fig. 5). The ROIs give the fit points for the functions: the center of the respective bin gives the coordinates of the point, and ROI characteristics, namely the size, the position and the rotation, give the point values. An example fit of one of the parameterization functions, $x_{\text{size}}(R_3)$ for the ROI in plane 2, is shown in Fig. 6.

2.4 Parameterization of track parameters

Similarly, to the parameterization of the ROIs, also the track parameters, such as the position and the momentum can be parameterized. It turns out that for parameterizing the momentum it is sufficient to

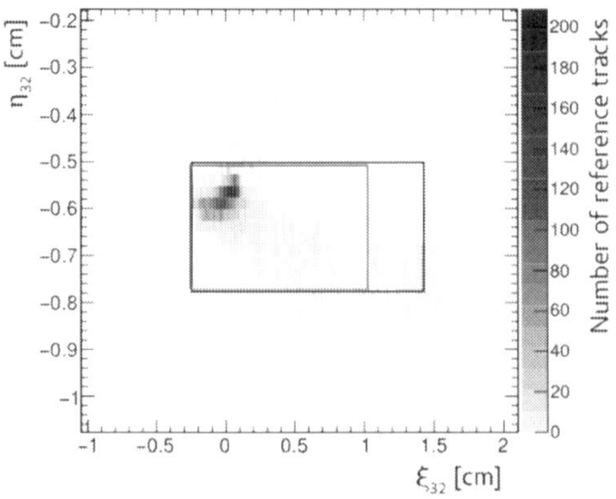

Figure 5. Spatial distribution of the reference tracks in plane 2 w.r.t. to the hit in plane 3 in the R_3 bin [73.5; 74.5) cm. The coordinates are explained in Fig. 4, as well as in the text. Black rectangle is the ROI, evaluated from this spatial distribution. Red rectangle is the ROI, evaluated from the parameterizations.

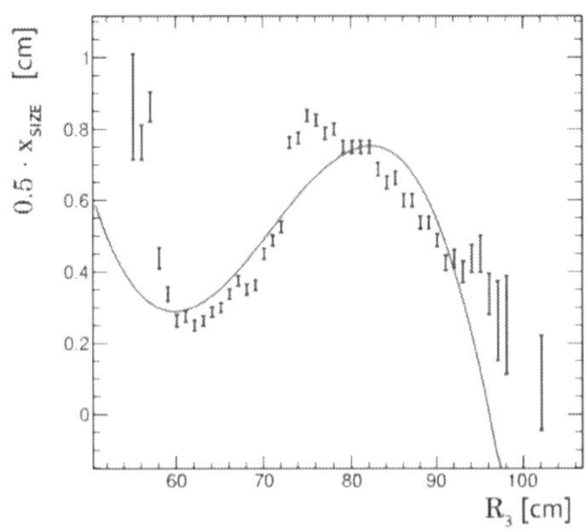

Figure 6. Fit of the $x_{size}(R_3)$ of the ROIs in plane 2 with a 3rd order polynomial. The size of the bars is inversely proportional to the weight of the point, which is defined as the number of reference tracks in the bin.

consider the positions of two hits only (necessarily from different plane pairs). Indeed, the curvature of the track is additionally constrained by the position of the target, and the other two hits are strongly correlated with the selected two. For the P2 experiment so far only the parameterization of the absolute momentum was implemented. It was chosen to use the hits in planes 1 and 3, and the variables R_3, ΔR_{31}, $\Delta\phi_{31}$. The results are presented in section 3.3. However, in order to completely replace the fitting procedure one would need to parameterize the χ^2 of fit, which necessarily requires considering the positions of all four hits. Whereas using 7 independent variables (two coordinates per hit minus one rotational symmetry) is clearly not possible in the way described above, it might be possible to either use a sophisticated function(s) of the hit coordinates as an argument for the parameterization function, or to develop (adopt) an appropriate method of evaluating the coefficients of the parameterization functions, which does not require a 7-dimensional binning. Further studies in this direction have not yet been done, but are planned.

3 Performance tests

A series of tests of the described tracking technique has been performed on the GEANT4-simualted data. Two different data samples were used to create the parameterizations, and to evaluate the performance. The samples were created with exactly the same GEANT4 model, as briefly described in the section 1.5.

The parameterizations were created using the reference tracks, reconstructed from simulated 45-ns time frames at 0.1% of the full beam rate. All reasonable four-hit combinations were constructed and fitted with the General Broken Lines (GBL) method ([6]), and the best tracks were selected by χ^2.

3.1 Efficiency of the parameterized regions of interest

The efficiency of a parameterized ROI is defined as the ratio of the number of signal tracks, that are within the ROI, to the total number of reconstructible signal tracks. A signal track is considered reconstructible if it made a hit in each of the four tracking planes. Only electrons that hit the Cherenkov Detetor are considered as signal. A more illustrative measure of the performance of a parameterized ROI is the distribution of the position of the signal track w.r.t. the center of the ROI. Both measures of ROI performance were evaluated with the current, yet not optimized, configuration of the algorithm.

A set of reconstructible signal tracks was selected, and for each track the ROIs in planes 2, 1 and 0 were constructed. The bottom right bar chart in Fig. 7 shows the fraction of signal tracks within individual ROIs, and ROI combinations. The overall efficiency of constructing track candidates is given by the bin "123", and equals 91 % in the given case. The inefficiency is caused partially by imperfect parameterization with 3rd order polynomials, and partially by hard bremsstrahlung by the signal electrons.

The three histograms in Fig. 7 show the distribution of the position of the signal track w.r.t. the center of the ROI. The quantity is plotted in the ROI local reference frame (Fig. 4), and in the units of ROI half-size. In such coordinates the tracks inside $-1 < x < 1 \bigcap -1 < y < 1$ are within the ROI.

This test proves the concept of parameterizing the ROIs. Further studies are needed to evaluate the perfomance of this technique.

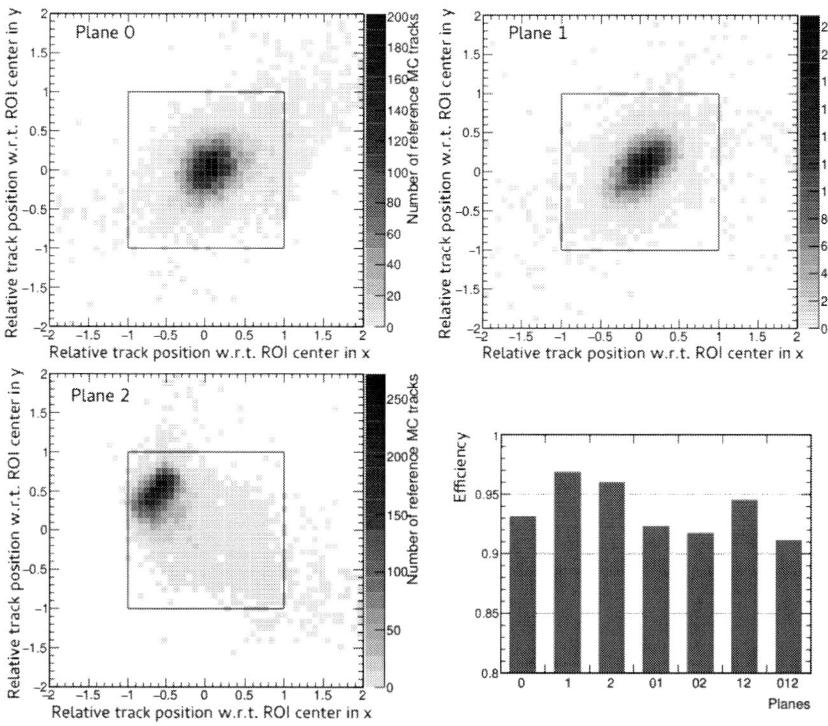

Figure 7. Top left, top right, bottom left: Position of test tracks w.r.t. to the center of the ROI, evaluated in the ROI local reference frame, and expressed in units of ROI half-size. The red squares indicate the ROI boundaries. Bottom right: efficiencies of ROI, and their combinations (explained in the text).

3.2 Combinatorial background

To see whether using parameterized ROIs brings any advantages in performance, a comparison to an algorithm using constant-size ROIs was done. The constant-size ROIs were chosen to be square, their size in planes 0 and 2 was set equal, and the size was tuned to the same efficiency (91%) as was obtained with the parameterized ROIs. The ROI in plane 1 was made so large that all hits from plane 1 were always enclosed (actually, it would be enough to select hits only from the same sector, and not from the whole plane, which would result in factor 4 fewer combinations).

Fig. 8 shows the number of constructed track candidates per one reconstructible signal track using the constant-size and the parameterized ROIs. Clearly, the difference is dramatic. One can expect that after fitting and applying a χ^2 cut most of the wrong candidates, constructed using the fixed-size ROIs, will be rejected, and the amount of combinatorial background in the two cases will become very similar. The advantage of using parameterized ROIs is that it eliminates the need to perform the huge number of unnecessary fits.

One has to keep in mind that the parameterization-based tracking implicitly uses the information about the kinematic distribution of the true tracks, whereas the algorithm with the fixed-size ROIs doesn't.

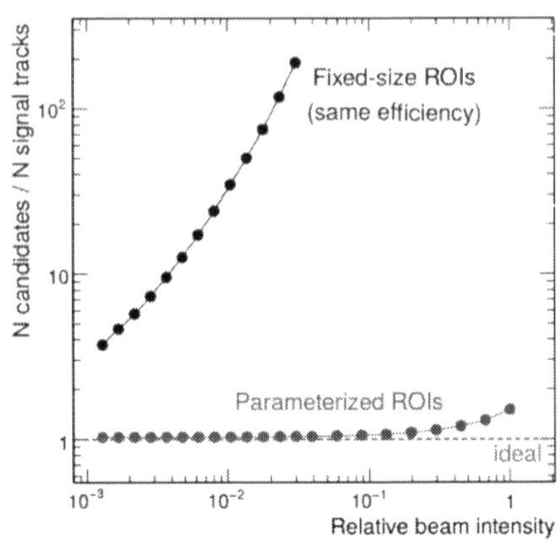

Figure 8. Number of track candidates, constructed per one reconstructible signal track, using the parameterized ROIs, and using the constant-size ROIs, tuned to the same efficiency.

3.3 Parameterization of track momentum

Following section 2.4 the absolute momentum was also parameterized. The accuracy of the parameterization was evaluated on the same sample of test tracks, as was used in sections 3.1 and 3.2. The obtained distribution of the relative error on the momentum is shown in Fig. 9. For comparison, the same distribution for the momentum, determined with a rigorous GBL fit is adduced. The parameterization provides a very good estimate for the track momentum. The difference in the quoted widths should rather be attributed to the imperfect Gaussian fits. The cause of the difference in the tails of the distributions is to be clarified.

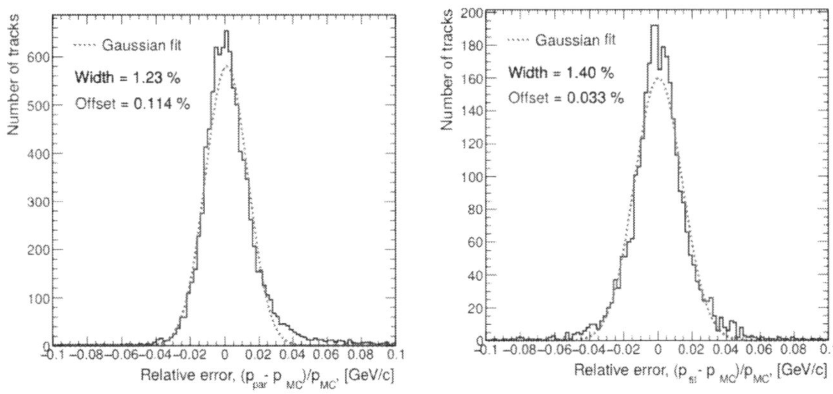

Figure 9. Relative error on the absolute momentum, evaluated from the parameterization (left), and determined with a rigorous GBL fit (right).

4 Summary

The parameterization-based tracking is a variation of track following, where the regions of interest are pre-calculated, and represented as a set of parametric analytic functions of the hit coordinates. In the current implementation the ROIs are defined by comparing the track segment with the reference tracks. Also the track parameters, such as the position and the momentum can be parameterized.

The parameterization-based tracking has been implemented for the P2 experiment and tested on simulated data. The first tests have proven the concept, and demonstrated a great improvement in the number of constructed track candidates, hence required fits, as compared to an algorithm with fixed-size ROIs. The parameterization of the absolute momentum is almost as accurate as the rigorous fit.

References

[1] N. Berger *et al.*, J. Univ. Sci. Tech. China **46**, no.6, 481-487 (2016)
[2] K. Aulenbacher, AIP Conf. Proc. **1563**, 5-12 (2013).
[3] R. Bucoveanu, M. Gorchtein, H. Spiesberger PoS LL2016, 061, (2016)
[4] H. Augustin *et al.*, Nucl.Instrum.Meth. **A845**, 194-198 (2017)
[5] J. Allison *et al.* Nucl.Instrum.Meth. **A835**, 186-225 (2016)
[6] C. Kleinwort, Nucl.Instrum.Meth. **A673** 107-110 (2012)
[7] A.-K. Perrevoort for the Mu3e collaboration EPJ Web Conf. **118** 1028 (2016)
[8] H. Augustin *et al.*, JINST 12, no.01, C01087 (2017)

Online Data Reduction using Track and Vertex Reconstruction on GPUs for the Mu3e Experiment

Dorothea **vom Bruch**[1,a] for the Mu3e collaboration

[1] *Institut für Kernphysik, Johann-Joachim-Becher-Weg 45, Johannes Gutenberg-Universität Mainz, 55128 Mainz, Germany*

Abstract. The Mu3e experiment searches for the lepton flavour violating decay $\mu^+ \rightarrow e^+e^-e^+$, aiming to achieve a sensitivity of $2 \cdot 10^{-15}$ in its first phase and ultimately aspiring to a final sensitivity of 10^{-16}. During the first phase of the experiment, a muon rate of $\sim 10^8\ \mu$/s will be available, resulting in a data rate of $\sim$ 80 Gbit/s. The trigger-less readout system is based on optical links and switching FPGAs sending the complete detector data for a time slice to one node of the filter farm. A full online reconstruction is necessary to reduce the data rate to a manageable amount to be written to disk. Graphics processing units (GPUs) are used to fit tracks with a non-iterative 3D tracking algorithm for multiple scattering dominated resolution. In addition, a three track vertex selection is performed by calculating the vertex position from the intersections of the tracks. Together with kinematic cuts, this allows for a reduction of the output data rate to below 100 MB/s using 12 DAQ PCs.

1 The Mu3e Experiment

The Mu3e experiment [1] is designed to search for the lepton flavour violating decay $\mu^+ \rightarrow e^+e^-e^+$. In the Standard Model (SM) this process is only allowed via neutrino mixing in loops, and it is heavily suppressed to below a branching fraction of 10^{-54} [2]. Therefore any observation of lepton flavour violation in the charged lepton sector is a clear indication for new physics. Various models beyond the SM predict charged lepton flavour violation at a level to which future detectors are sensitive. The current limit on the $\mu^+ \rightarrow e^+e^-e^+$ branching fraction was set by the SINDRUM experiment at 10^{-12} [3]. The Mu3e experiment aims to reach a single event sensitivity of $2 \cdot 10^{-15}$ in a first phase of the experiment with an existing beamline at the Paul Scherrer Institute in Switzerland (PSI), and aspires to a final sensitivity of 10^{-16} with an upgraded beamline. This will improve the limit by four orders of magnitude compared to the last experiment. In these proceedings, the focus is on the experimental setup planned for the existing beamline.

The detector design is driven by the requirements to distinguish the signal decay from background processes. Within the detector volume, muons will be stopped in a target and decay at rest. In the case of a signal event, two positrons and one electron are coincident in time and originate from one single vertex as shown in figure 1a. The combined energy of the three particles is equal to the rest mass of the muon and their combined momentum is zero. One source of background is radiative muon decay

[a] e-mail: vombruch@uni-mainz.de

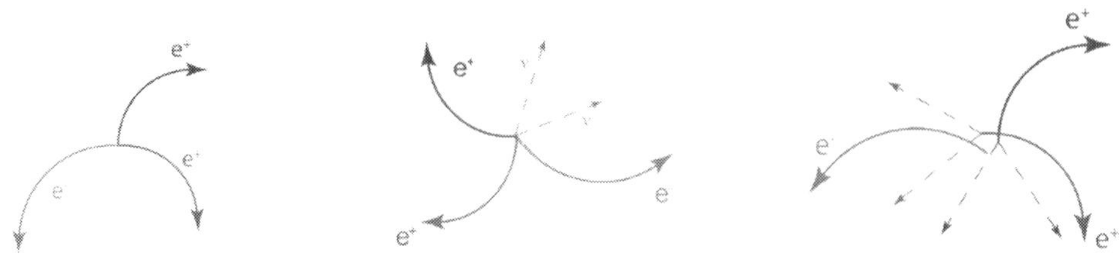

(a) Signal event (b) Internal conversion (c) Combinatorial background

Figure 1: Comparison of signal and background events. (a) $e^+e^-e^+$ signal decay particles from one single vertex. (b) Background from internal conversion in an ordinary muon decay ($\mu \to e^+\nu_e\bar{\nu}_\mu e^+e^-$), decay particles from one single vertex, but the neutrinos carry away some energy. (c) Combinatorial background due to positrons from ordinary muon decay combined with electrons produced by Bhabha scattering, photon conversion etc., decay particles not from a single vertex.

where the photon undergoes internal conversion into an e^+e^- pair: $\mu^+ \to e^+e^-e^+\bar{\nu}_\mu\nu_e$ as shown in figure 1b. In this case, the detectable decay products are also coincident in time and have a single vertex, but their momenta do not add up to zero and the combined energy does not equal the muon rest mass. A second class of background process stems from the combination of two ordinary muon decays $\mu^+ \to e^+\nu_e\bar{\nu}_\mu$ close to one another spatially and temporally together with an additional electron from photon conversion, Bhabha scattering etc. (see figure 1c). When both the e^+ and the e^- from Bhabha scattering or photon conversion are detected, only one additional muon decay is required and the probability of misreconstruction as a signal event is higher.

In order to discriminate signal events from background, excellent momentum, timing and vertex resolution are required. To achieve the desired sensitivity, Mu3e aims for a momentum resolution <0.5 MeV/c, a timing resolution of 100 ps and a vertex resolution of <200 µm. The Mu3e detector was designed to meet all of these requirements.

1.1 Detector Design

The momentum of the positrons from muons decaying at rest is at maximum half the muon rest mass (53 MeV/c). For these low momentum positrons, multiple Coulomb scattering is the main contribution to the momentum resolution, as its variance is inversely proportional to the momentum. Consequently, the Mu3e detector components are aimed at minimising the amount of material traversed. High Voltage Monolithic Active Pixels Sensors [4, 5] (HV-MAPS) thinned down to 50 µm with a pixel size of 80×80 µm^2 placed on ultralight mechanics [6] are used for tracking. In the central part of the detector, two layers of HV-MAPS separated by 6 mm closely surround the hollow double cone target to measure the trajectory's direction for a good vertex reconstruction. An additional doublet of layers is located at larger radius for a second direction measurement, providing a momentum measurement from the particle's curvature in the magnetic field. A schematic drawing of the detector is shown in figure 2. On either side of the central part, recurl stations consisting of another two tracking layers measure the trajectories of particles recurling in the 1 T magnetic field, much improving the momentum measurement. Just inside the third layer of pixel detectors, plastic scintillating fibres are included in the central part and plastic scintillating tiles are positioned in the recurl stations for precise timing measurements. With three layers of fibres, a time resolution of 550 ps was obtained, the scintillating

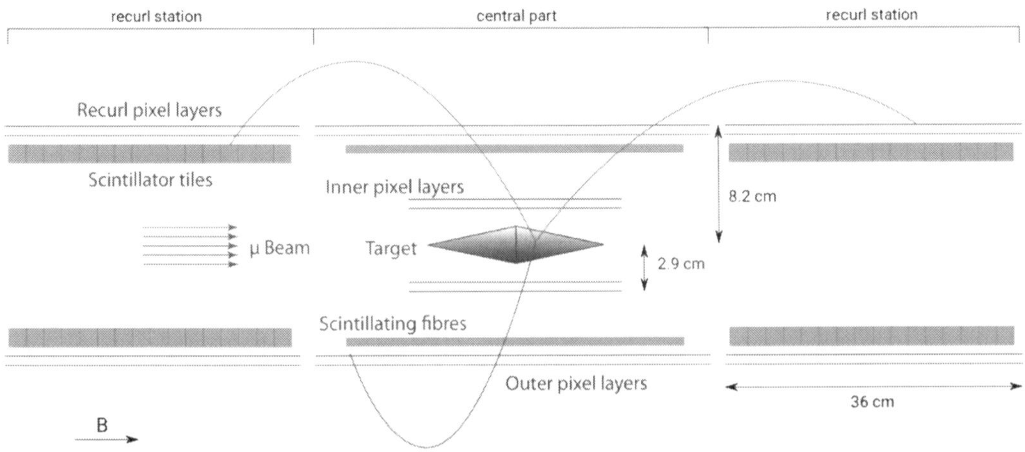

Figure 2: Schematic drawing of the Mu3e detector design.

tiles have a resolution of 70 ps. Muons at high rates are desired to reach the projected sensitivities within a reasonable amount of time. Current beamlines at PSI deliver up to $1 \cdot 10^8\,\mu$/s. A future high intensity muon beamline could deliver more than $2 \cdot 10^9\,\mu$/s.

1.2 Readout Scheme

The detector employs a trigger-less readout architecture. At a muon stopping rate of $1 \cdot 10^8\,\mu$/s this results in a data rate of about 80 Gbit/s. About 100 Mbyte/s can be written to disk, which requires an online system for selecting signal decays. A schematic drawing of the data acquisition is shown in figure 3. The pixel sensors, the fibres and the tiles send zero-suppressed data to front-end Field Programmable Gate Arrays (FPGAs) via flex print cables. Here, the data from a part of the detector is merged and sorted into time slices of 50 ns. These are transferred via optical links to the switching boards where the time slices from each sub-component of the detector are merged. The complete detector information for one time slice is delivered to a single computer of the filter farm via optical links. It is received by an FPGA and shipped to the main memory of the PC by direct memory access (DMA) over a PCIe connection. When analysing 50 ns time slices, the 12 DAQ computers need to process $2 \cdot 10^7$ slices / s, resulting in $1.7 \cdot 10^6$ slices/s each. In order to manage the computing load, Graphics Processing Units (GPUs) are used for the track fitting and vertex selection. The data relevant for the online selection is therefore transferred to each PC's GPU via DMA, where the selection process takes place.

2 Online Signal Selection

After a first selection step, tracks are reconstructed and classified as either electrons or positrons. In a second step, track intersections are searched for to select events with three tracks originating from one single vertex fulfilling the kinematic characteristics of a signal decay.

 DOI: 10.1051/epjconf/201715000013

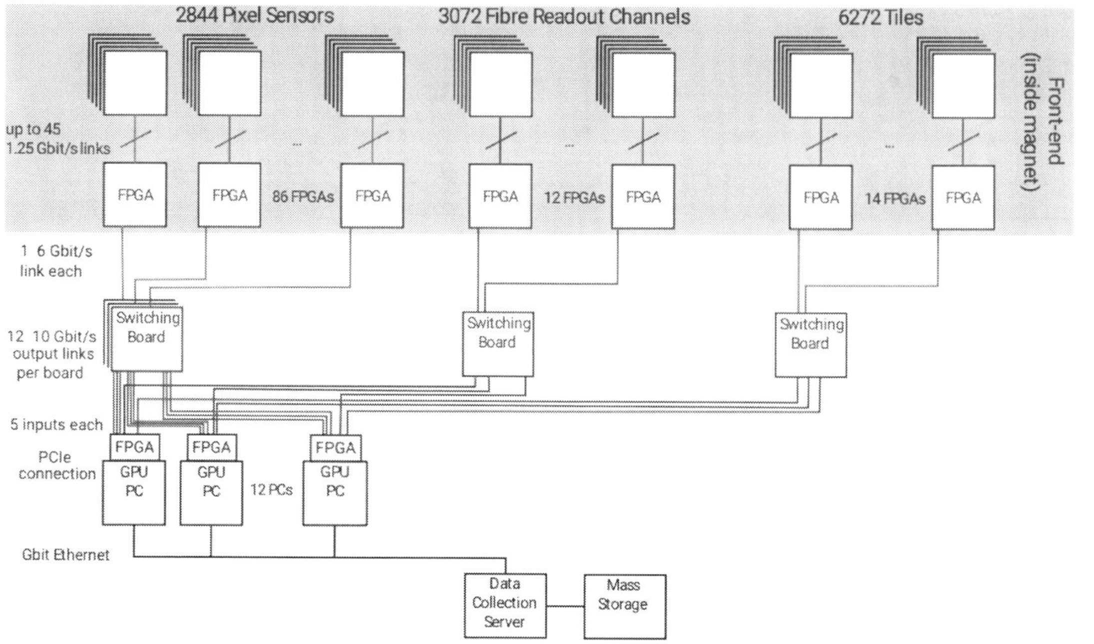

Figure 3: Readout scheme of the Mu3e detector.

2.1 Geometrical Selection

Before the actual track fit, combinations of hits in the first three detector layers are selected based on few simple geometrical selection criteria. The first is the difference in the z-coordinate of hits between subsequent layers. The second is the difference of the angles in the plane transverse to the beam direction. The mean number of hits per 50 ns time slice in the first and second doublet of layers is ~10, and ~13 respectively, so >1000 three-hit combinations are candidates for the track fit. After applying the geometric selection cuts for differences of hits between the first, second and third layers, this number is reduced by a factor of 70.

2.2 Track Fit

Hits from the central stations of the pixel detector are used to reconstruct tracks using a 3D tracking algorithm developed for multiple scattering dominated resolution [7]. Three hits in subsequent layers, called a "triplet", are selected and multiple scattering is assumed to occur at the middle hit of the triplet. The effect of multiple scattering dominates over the position resolution of the pixel detector ($\sigma_{\text{pixel}} = 80\,\mu\text{m} / \sqrt{12}$) in the momentum range of the Mu3e experiment, so the latter is ignored in the fit. For details on this novel multiple scattering fit see reference [8]. In the online track reconstruction, the first track estimate is obtained from a triplet of hits in the first three detector layers. This track is propagated to the fourth layer and the hit closest to the propagated position is chosen for a refit with a second triplet, whose first two hits are equal to the last two hits of the first triplet. For a sample of simulated events with particles interacting according to SM branching fractions, so that positrons originate mainly from ordinary muon decay ($\mu^+ \rightarrow e^+ \nu_e \bar{\nu}_\mu$) (from now on referred to as background sample), tracks are selected with a χ^2 cut keeping 97 % of true 4-hit tracks while 59 % of the sample

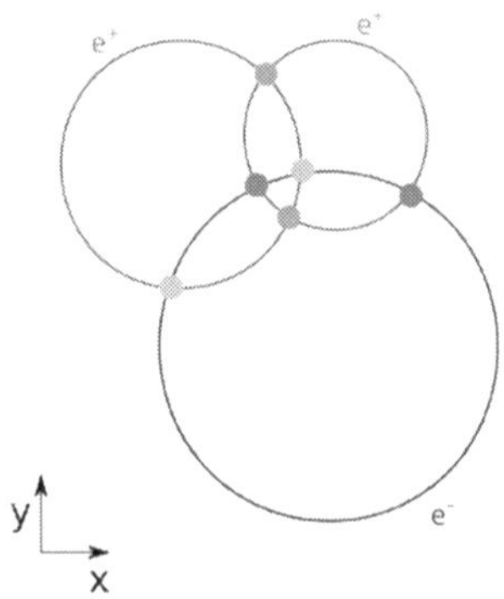

(a)

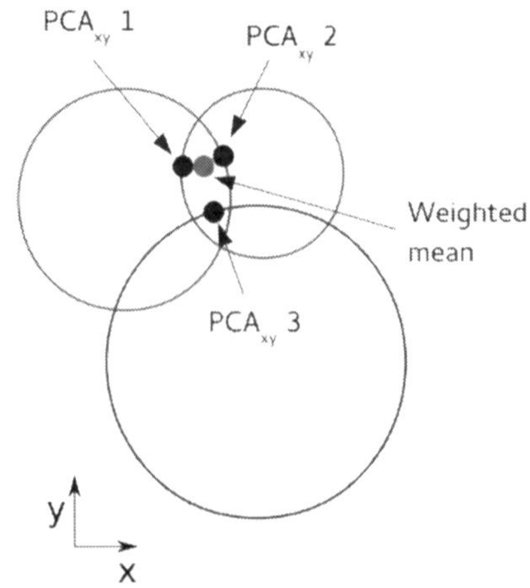

(b)

Figure 4: Combination of three tracks projected onto the plane transverse to the beam direction. (a) Intersections of two circles are shown in the same colour. (b) Weighted mean of three intersections from different tracks (different colours in (a)) and the points of closest approach.

are true tracks. The tracks are sorted according to their curvature in the magnetic field into positive and negative tracks to perform the vertex search for two positrons and one electron.

2.3 Vertex Estimate

The vertex search is based on simple geometric constraints to meet the stringent performance requirements of the online selection process. All combinations of two positrons and one electron are considered within each time slice of 50 ns. First, only the projection onto the plane transverse to the beam direction, where the helical tracks are circles, is studied to search for intersections of three circles (see figure 4a). For each intersection, weights are calculated based on the uncertainties due to multiple scattering in the first detector plane and due to the pixel size. If all three circles intersect, the weighted mean $\overline{xy}$ is calculated for all combinations of three intersections from three different tracks. For every $\overline{xy}$, for each track, the point of closest approach to $\overline{xy}$, PCA_{xy}, and its weight $\sigma_{PCA_{xy}}$ are determined (see figure 4b). At PCA_{xy}, the z-coordinate PCA_z of the track and also a weighted mean of the three z-coordinates $\overline{z}$ are calculated. The χ^2 is computed from the differences between the PCA and the weighted mean both in the transverse plane and in the z-coordinate:

$$\chi^2 = \sum_{i=1}^{3} \frac{\left|PCA_{xy,i} - \overline{xy}\right|^2}{\sigma^2_{PCA_{xy,i}}} + \frac{\left|PCA_{z,i} - \overline{z}\right|^2}{\sigma^2_{PCA_{z,i}}}$$

For each combination of three intersecting circles, the vertex estimate with the smallest χ^2 value is chosen. Figure 5 shows the χ^2 distribution for vertices originating from simulated events containing one $\mu^+ \rightarrow e^+e^-e^+$ vertex in very 50 ns time slice (from now on referred to as signal sample), and for

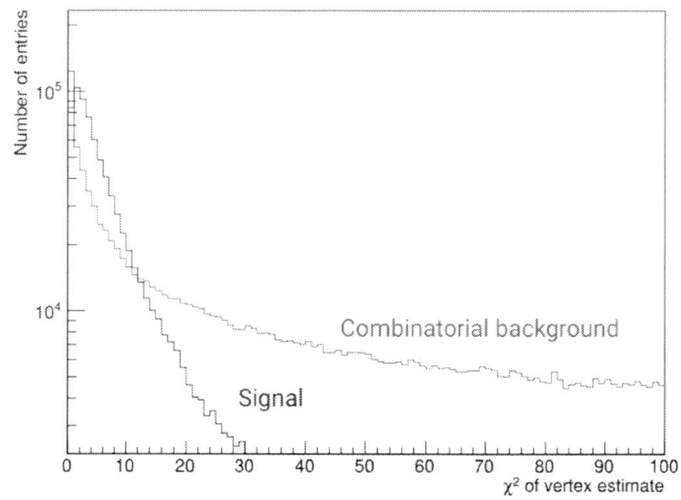

Figure 5: χ^2 distribution for vertices from true signal decays (blue) and from combinatorial background from two positron tracks and one electron track intersecting with one another near the target, from a background sample (red). Due to this preselection, the combinatorial brackground also peaks at 0. The normalisation scale is chosen arbitrarily for visualisation purposes.

combinatorial background from three tracks intersecting near the target from a simulated background sample. A vertex resolution of ~440 µm in the transverse plane and ~300 µm in the z-coordinate is achieved. For the selection of signal decays, cuts on the χ^2 value and on the distance between the estimated vertex position and the target surface are applied. In addition, the combined momentum and energy are reconstructed for the three tracks and decays at rest with an energy of the muon rest mass are selected by kinematic cuts. Finally, fake vertices produced by tracks recurling in the central part of the detector are reduced by requiring a minimum opening angle for tracks with small momentum difference.

2.4 Signal Efficiency and Accepted Background Fraction

The accepted background fraction is simply given by the fraction of time slices accepted by the selection cuts when simulated background events are used as input. To estimate the signal efficiency, simulated signal events were studied, whose tracks were reconstructed by the offline reconstruction framework which also takes into account hits from the recurl pixel stations and reconstructs long recurling tracks with six and eight hits. Afterwards, a linearized vertex fit [9] is performed and cuts are applied on the χ^2 of the fit, the distance to the target and the combined momentum magnitude. Figure 6 shows the number of signal time slices selected by both the online and offline selection processes normalised to the number of signal time slices selected by the offline selection chain, as well as the accepted background fraction for the different online selection cuts. After all cuts, 98 % of signal time slices remain and the fraction of accepted background time slices is reduced to 0.7 %. This delivers more than the factor 100 reduction in data rate required from the online selection process.

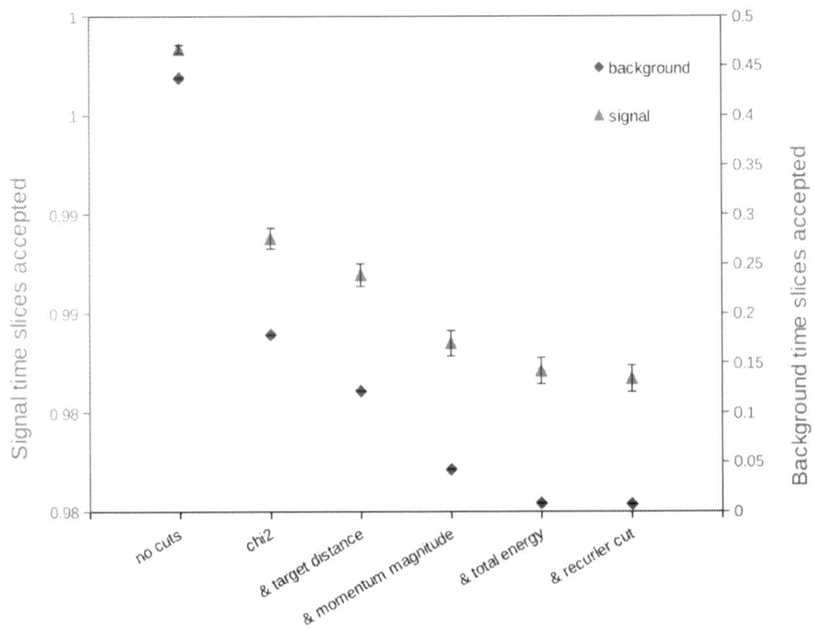

Figure 6: Fraction of signal (blue triangles) and background (red diamonds) time slices for the different online selection cuts described in the text. "No cuts" means that no vertex quality cuts are applied, however, the time slice is required to contain two positrons and one electron with 2-track intersections close to the target.

3 Implementation on GPUs

The geometrical preselection will be carried out on the FPGA connected to the DAQ PC via PCIe connection. The track fit and vertex selection have been implemented in CUDA to run on Nvidia GPUs. The triplets of hits selected by the preselection on the FPGA are written into a memory layout suitable to make full use of the memory bandwidth provided by the GPU when accessing the information for the track fit.

The following usage of the hardware was optimised for a GTX1080Ti. This device has 3584 compute units available, the main challenge consists in distributing the work of track fitting and vertex selection on them. To this end, for 24×8192 time slices the selection procedure is started in parallel. For each of the time slices, one block of 128 threads performs the calculations, each thread doing the track fit for one combination of three hits, and looping over the hits in the fourth layer. After all track candidates have been fitted, each of the same 128 threads is used to investigate combinations of one electron and one positron. In each thread, a loop over the positrons in this time slice is performed and the selection decision is taken according to the vertex selection process described in section 2.3.

In general, most threads on a GPU are scheduled independently of each other. However, within one sub-unit of threads, a so-called block, it is possible to synchronise the individual threads and a shared memory region with fast access is available. Therefore, one block was chosen per time slice, so that the track parameters can be stored in the shared memory and the synchronisation can be done after the track fitting step and before the vertex selection to ensure that all tracks have been found before the vertex selection begins.

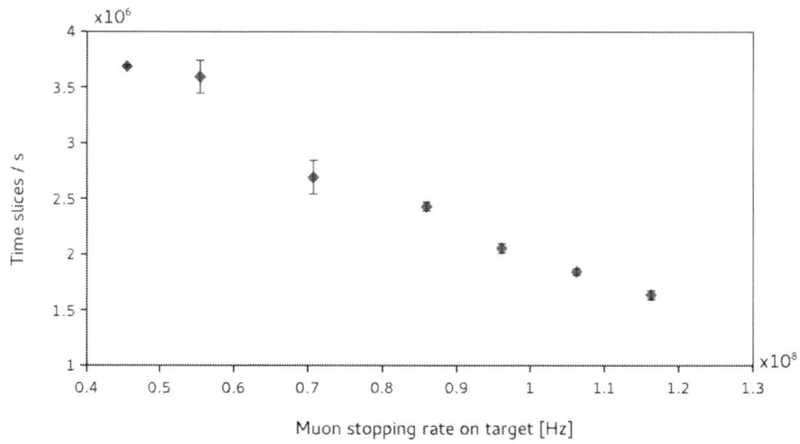

Figure 7: Performance of the online even selection algorithm running on a single Nvidia GTX1080Ti GPU versus the muon stopping rate on target.

In addition to an efficient distribution of the workload, it is critical that each single thread performs the selection as quickly as possible. To this end, optimisation techniques known for CPU code were applied as well. For example, values are only saved for reuse if a recomputation takes longer than storing the result in a register. Also, exit conditions of loops are ordered such that the most probable case comes first and trigonometric functions are avoided if possible.

To limit the amount of memory needed on the GPU and to prevent long computation durations for single time slices with high combinatorics, for each time slice the number of hits per layer, the number of three-hit combinations accepted by the geometrical preselection and the number of reconstructed positron and electron tracks are limited. Any time slices exceeding these limits are stored to disk and contribute to the background fraction.

After the optimisations described above, one Nvidia GTX1080Ti can process $2 \cdot 10^6$ time slices/s for events simulated at a muon stopping rate of $1 \cdot 10^8$ Hz, the rate expected during the first phase of the experiment. Figure 7 shows the strong dependence of the performance versus the muon stopping rate on target. The achieved number of time slices per second is sufficient for running with a filter farm of only 12 DAQ PCs. In addition, the filter farm will only be needed in roughly two years when the experiment will be commissioned. Until then, two more generations of Nvidia GPUs will become available on the market. In recent years, every new generation has increased the algorithm's performance by about a factor of three. This leaves enough margin for the performance presented in these proceedings and opens up possibilities to add more features to the online selection process.

4 Conclusions

During the first phase of the Mu3e experiment, the aim is to achieve a sensitivity of $2 \cdot 10^{-15}$ in the search for the decay $\mu^+ \rightarrow e^+e^-e^+$. The beamline will deliver $\sim 1 \cdot 10^8\,\mu$/s, resulting in a data rate of $\sim$ 80 Gbit/s. In order to reduce this data rate, an online selection process was implemented on GPUs. Tracks are fitted from hits in the central part of the detector and vertices originating from two positrons and one electron are searched for. This selection procedure was optimised to keep 98 % of the signal time slices and to reduce the background time slices to below 1 %. Due to the significant background reduction, the remaining data can be stored to disk. The selection chain was implemented on an Nvidia GTX1080Ti and its performance was optimised to process $2 \cdot 10^6$ time slices/s. This is sufficient when using the 12 DAQ PCs planned for the first phase of the experiment.

References

[1] A. Blondel, A. Bravar, M. Pohl, S. Bachmann, N. Berger, M. Kiehn, A. Schöning, D. Wiedner, B. Windelband, P. Eckert et al., Research Proposal to PSI (2012)

[2] W.J. Marciano, T. Mori, J.M. Roney, Annual Review of Nuclear and Particle Science **58**, 315 (2008)

[3] W. Bertl, S. Egli, R. Eichler, R. Engfer, L. Felawka, C. Grab, E. Hermes, N. Kraus, N. Lordong, J. Martino et al., (SINDRUM collaboration), Nuclear Physics B260 pp. 1–31 (1985)

[4] H. Augustin, N. Berger, S. Dittmeier, J. Hammerich, U. Hartenstein, Q. Huang, L. Huth, D. Immig, A. Kozlinskiy, F.M. Aeschbacher et al., Journal of Instrumentation **11**, C11029 (2016), `arXiv:1610.02210v2`

[5] H. Augustin, N. Berger, S. Dittmeier, C. Grzesik, J. Hammerich, Q. Huang, L. Huth, M. Kiehn, A. Kozlinskiy, F.M. Aeschbacher et al., NIM A **845**, 194 (2017), `arXiv:1603.08751`

[6] N. Berger, S. Dittmeier, L. Henkelmann, A. Herkert, F.M. Aeschbacher, Y.W. Ng, L.O.S. Noehte, A. Schöning, D. Wiedner, Journal of Instrumentation **11**, C12006 (2016), `arXiv:1610.02021v1`

[7] N. Berger, A. Kozlinskiy, M. Kiehn, A. Schöning, NIM A **844**, 135 (2017), `arXiv:1606.04990`

[8] A. Kozlinskiy, in these proceedings (2017)

[9] S. Schenk, Bachelor thesis, Heidelberg University (2013), `https://www.psi.ch/mu3e/ThesesEN/BachelorSchenk.pdf`

Online Data Reduction for the Belle II Experiment using DATCON

Florian **Bernlochner**[1], Bruno **Deschamps**[1], Jochen **Dingfelder**[1], Carlos **Marinas**[1], and Christian **Wessel**[1,a]

[1] *Physikalisches Institut, University of Bonn, Germany*

Abstract. The new Belle II experiment at the asymmetric e^+e^-accelerator SuperKEKB at KEK in Japan is designed to deliver a peak luminosity of $8 \times 10^{35}\text{cm}^{-2}\text{s}^{-1}$. To perform high-precision track reconstruction, e.g. for measurements of time-dependent CP-violating decays and secondary vertices, the Belle II detector is equipped with a highly segmented pixel detector (PXD). The high instantaneous luminosity and short bunch crossing times result in a large stream of data in the PXD, which needs to be significantly reduced for offline storage. The data reduction is performed using an FPGA-based Data Acquisition Tracking and Concentrator Online Node (DATCON), which uses information from the Belle II silicon strip vertex detector (SVD) surrounding the PXD to carry out online track reconstruction, extrapolation to the PXD, and Region of Interest (ROI) determination on the PXD. The data stream is reduced by a factor of ten with an ROI finding efficiency of >90% for PXD hits inside the ROI down to 50 MeV in $p_T$of the stable particles.

We will present the current status of the implementation of the track reconstruction using Hough transformations, and the results obtained for simulated $\Upsilon(4S) \to B\bar{B}$ events.

1 Motivation for the Belle II Experiment

Since the Belle experiment stopped taking data in 2010, its successor Belle II is being developed. The Belle II experiment will be located at the KEK laboratory in Japan, at the only interaction point of the SuperKEKB collider, which is the successor of the KEKB machine. SuperKEKB will collide asymmetric energy beams (7 GeV electrons and 4 GeV positrons) with a design instantaneous luminosity of $8 \times 10^{35}\text{cm}^{-2}\text{s}^{-1}$. This luminosity is 40 times the instantaneous luminosity of KEKB. This produces higher data rates and thus higher background rates with which the new Belle II detector has to cope. Due to limited bandwidth of the readout electronics and to minimise the amount of mass storage required, online data reduction is essential. The working principle of one of the two deployed online data reduction systems, the Data Acquisition Tracking and Concentrator Online Node (DATCON) is described in this article.

2 The Belle II Vertex Detector

A sketch of the Belle II detector is shown in Figure 1. The innermost part of the detector is the silicon based vertex detector system (VXD), which is surrounding the beryllium beam pipe. The other

[a] e-mail: wessel@physik.uni-bonn.de

DOI: 10.1051/epjconf/201715000014

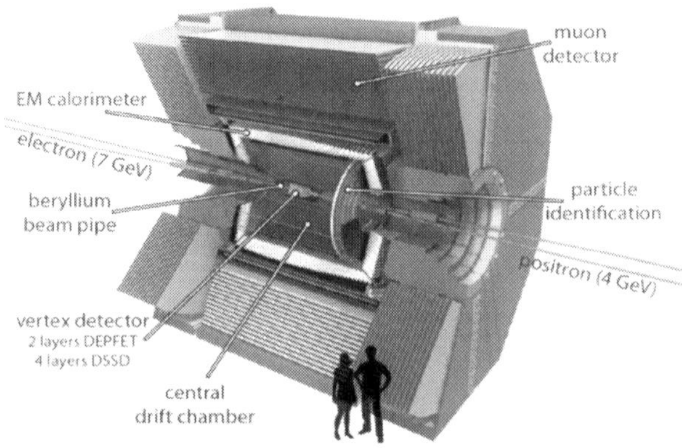

Figure 1. Overview of the Belle II Detector.

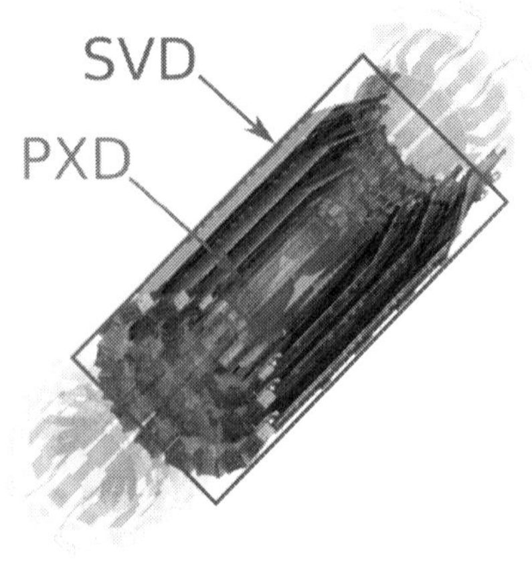

Figure 2. The Belle II Vertex Detector. It contains two subdetectors, the PiXel Detector (PXD) with two layers, surrounded by the Silicon Vertex Detector (SVD) with four layers. The innermost radius is 14 mm, the outermost radius is 135 mm.

sub-detectors are: the central drift chamber, the solenoid magnet producing a magnetic field of 1.5T, detectors for particle identification, the electromagnetic calorimeter, and an instrumented flux return that can detect K_L and muons. The VXD consists of two components: a PiXel Detector (PXD) based on DEPFET [1] technology and a Silicon Vertex Detector (SVD) based on double-sided silicon strip sensors. The placement of both components is shown in Figure 2. Both PXD and SVD are very thin detectors with only 0.2% and 0.6% of a radiation length per layer, respectively. The PXD consists of 2 layers at radii of 14 and 22 mm from the interaction point with in total 40 modules. A module contains 250 × 768 pixels, having two types of pixels with a pitch of 50 × 55 μm^2 (256 pixels in central region) and 50 × 70 μm^2 (512 pixels in forward and backward region) for layer 1 and 50 × 65 μm^2 (256 pixels in central region) and 50 × 70 μm^2 (512 pixels in forward and backward region) for layer 2. The information in the 8 million pixels is read out at a data rate of 256 Gb/s, corresponding to 90% of the data rate of the complete Belle II detector without data reduction. The SVD consists of 4 layers at radii ranging between 39 and 135 mm. In total the SVD contains 172 sensors with 768 × 768 strips having a pitch of 50 and 160 μm (layer 3) and 768 × 512 strips having a pitch of 75 × 240 μm (layers 4 to 6), respectively. This results in about 240000 strips in total [2].

In a $B\bar{B}$ event coming from the decay of the $\Upsilon(4S)$ resonance there are on average 10 tracks in the acceptance region of the VXD, which covers the full angle of 2π in azimuthal direction (φ), and the region $17° < \theta < 150°$ in polar angle. In addition to hits from particle tracks originating from $B\bar{B}$ events, background processes produce additional hits that are recorded by the VXD. Such background hits come for instance from two-photon QED processes, Touschek scattering, Coulomb scattering of beam particles with residual gas in the beam pipe, radiative Bhabha scattering and Synchrotron radiation. Beam-induced backgrounds increase the occupancy of the PXD up to 1% and up to 0.25% in the SVD, based on detailed simulations of the backgrounds. Only a small fraction of the hits in the VXD belong to tracks originating from the decay of $B\bar{B}$ pairs. The online data reduction of the Belle

II experiment is designed in such a way that only the hits of interest for physics analysis recorded in the PXD are forwarded to permanent storage. To implement this, the hits of other sub-detectors as the SVD are used to execute an online track finding. The reconstructed tracks are then used to define Regions of Interest (ROI) on the PXD and only the subset of pixels inside an ROI are permanently stored. For the maximum tolerable occupancy of 3%, a data reduction factor of about 10 for PXD hit information is required.

3 Data Reduction Concept

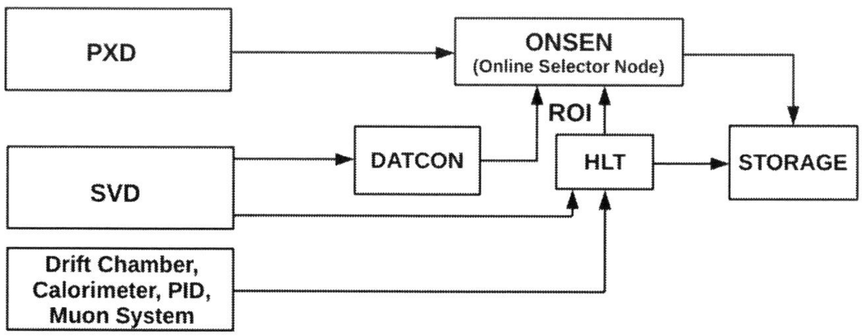

Figure 3. A simplified illustration of the Belle II data acquisition system with the two data reduction systems: DATCON receives hit recorded by the SVD and performs an online track reconstruction to define ROIs on the PXD. The HLT receives hit information from the SVD, the drift chamber, and the muon system as well as information from the PID detectors to define ROIs on the PXD. Both HLT and DATCON run independently of each other. The ROIs of both systems are sent to the ONSEN system, which performs the overall PXD data reduction by applying the ROIs to the PXD data.

Figure 3 shows a simplified sketch of the Belle II Data Acquisition (DAQ) system. The data of the SVD, the drift chamber and the sub-detectors for particle identification (PID), calorimetry and muon detection are sent to the Higher Level Trigger (HLT) [3]. The data from the SVD are additionally sent to the DATCON, which performs online track reconstruction, extrapolation to the PXD, and calculation of ROI's in the PXD. The ROIs are then sent to the Online Selector Node (ONSEN) [4]. Besides DATCON, the HTL is the second system that performs a calculation of ROI's in the PXD. For this task the HLT not only can use data from the SVD, but also from the drift chamber and the other sub detectors. In addition, HLT provides the trigger signal for the complete detector and pipelines the data of the sub-detectors, except PXD, to the storage. The PXD data are only sent to ONSEN, which merges the ROI's of HLT and DATCON and performs the overall PXD data reduction by rejecting hits outside the ROIs. The HLT uses a computing farm with 6400 cores in total and runs sophisticated track finding and fitting algorithms. These HLT algorithms will also be used in the later offline track reconstruction. DATCON, on the other hand, runs a fast FPGA-based track reconstruction.

A track in a magnetic field, parametrised by a helix, is characterised by five parameters at the point of closest approach from to the beamline. This is illustrated in Figure 4. The parameters are: the initial azimuthal angle φ_0, the distance of closest approach d_0, the radius of the track $r_{T,0}$, the initial polar angle θ_0, and the initial z-coordinate at the point of closest approach z_0.

A schematic overview of the data reduction procedure and the data flow inside DATCON is shown in Figure 5: The ROIs are calculated by using SVD hits to reconstruct tracks and extrapolating the

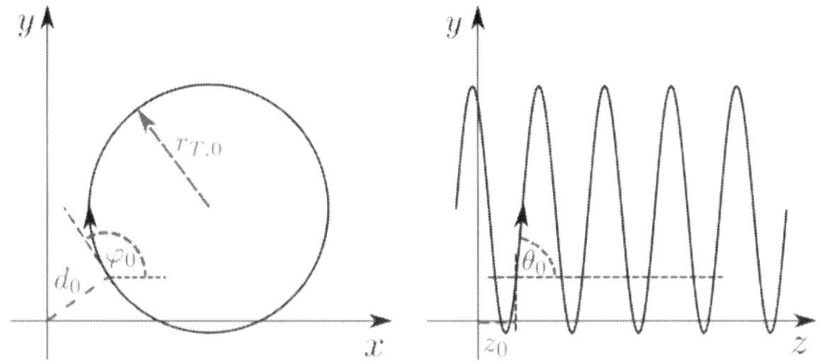

Figure 4. Two-dimensional view of a helix as parametrisation of a track inside a magnetic field. The index 0 indicates parameters at the point of closest approach of an ideal track. The values of these parameters can change with time due to e.g. multiple scattering or other processes which cause energy loss.

trajectories to the PXD. For the purpose of track finding, the tracks are assumed to be circular in the x-y-plane , with an additional ghost hit at the origin of the coordinate system where the beam spot is located. The algorithm assumes that the trajectories can be approximated by straight lines in the r-z-plane with $z_0 \neq 0$. The hits are transformed using Hough [5] and Hesse transformations:

$$\rho = \frac{2}{r_{\text{hit}}^2} \cdot \left(x' \cos\varphi + y' \sin\varphi\right), \tag{1}$$

$$s = r_{\text{hit}} \cos\theta + z_{\text{hit}} \sin\theta, \tag{2}$$

where $\rho = \frac{1}{r_T}$ denotes the track curvature, r_{hit} is the distance of the hit from the z-axis , z denotes the global z coordinate of the hit, and (x', y') are the conformally transformed values of (x, y) of the hit defined by:

$$\left(x', y'\right) = \frac{(x_{\text{hit}}, y_{\text{hit}})}{r_{\text{hit}}^2} = \frac{(x_{\text{hit}}, y_{\text{hit}})}{x_{\text{hit}}^2 + y_{\text{hit}}^2}. \tag{3}$$

The conformal transformation is only valid for $d_0 = 0$, which is a good approximation for B-meson decays as their decay products (except the charged particles $e^{\pm}, \mu^{\pm}, \pi^{\pm}$, and $p/\bar{p}$) only have a short life-time and decay in close proximity to the origin of the z-axis . Note that the conformal transformation is needed as the Hough transformation is better applicable to straight lines. Using conformal and Hough transformation, a helix trajectory is mapped onto a straight line. Hits on this straight line correspond to intersecting of lines in the Hough parameter space. Thus the task of finding tracks in real space is equivalent to finding intersections of lines in Hough space. Once the intersections are found, equations 1 and 2 allow for a straightforward computation of the angle and track curvature $\rho = \frac{1}{r_T}$.

With this information intersections of all tracks with the PXD detector planes are calculated. This reduces the task of finding ROIs to finding the intersection of a circle with a straight line in two dimensions in the x-y-plane and to finding the intersection of two straight lines in the r-z-plane . From these interceptions, Most Probable Hits (MPH) are defined and a fixed-size ROI of 80 × 120 pixels is created. Studies to tune the ROI size on the estimated particle momenta are currently being performed. Finally, the identified ROIs are transmitted to ONSEN.

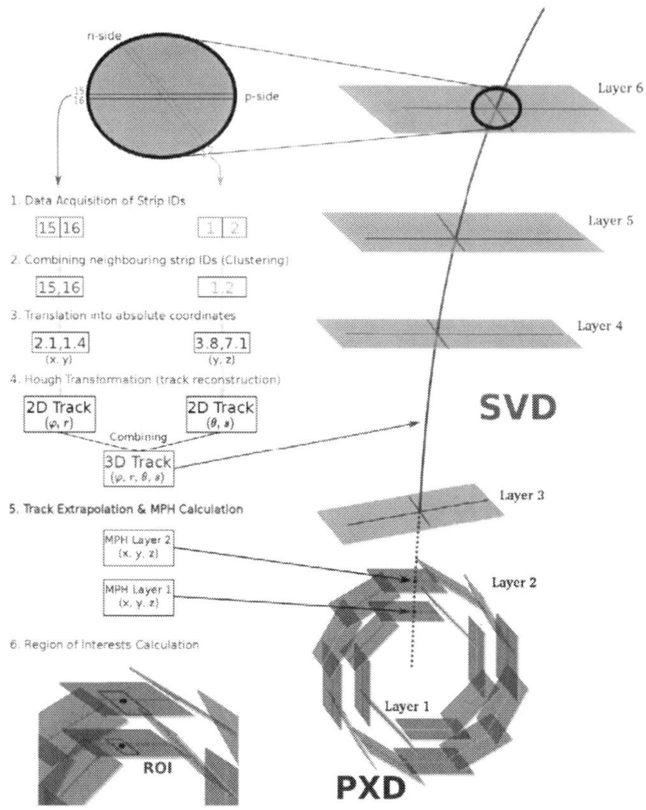

Figure 5. Schematic overview of the DATCON track reconstruction and extrapolation to the PXD. The data of the the SVD strips (1.) are converted to clusters (2.) in local detector coordinates, from which the absolute coordinates of the hits are computed (3.). These coordinates are Hough-transformed (4.) to reconstruct possible tracks, yielding two 2D tracks with information on φ and r, and θ and s, respectively. The information is combined to obtain 3D tracks, which are extrapolated to the PXD. The positions of the most probable hits are calculated from the intersection of the extrapolated tracks with the PXD layers (5.). Finally the ROI's on the two PXD layers are calculated (6.).

4 Preliminary Results

To develop and test the necessary algorithm to be implemented on an FPGA, a C++ and python based implementation of the algorithm is used, running inside the Belle II Analysis and Software Framework (BASF2) [6]. The BASF2 framework allows for the simulation of the detector response of particles traversing the Belle II detector and includes the full decay chain of $\Upsilon(4S) \rightarrow B\bar{B} \rightarrow$ stable particles (i.e. $e^{\pm}$, $\mu^{\pm}$, $\pi^{\pm}$, $p/\bar{p}$, and K_L). This allows one to assess the performance of a given algorithm using simulated decays and variable background conditions. The preliminary performance of such simulations of the DATCON using 100,000 simulated $B\bar{B}$ events are shown.

Figure 6 shows the performance of the track reconstruction for the azimuthal and the polar angle: 92% and 93% of all reconstructed tracks are within 1° of the true track in φ and θ, respectively.

In Figure 7, the track reconstruction efficiency as a function of the transverse momentum, p_{T}, is shown: The overall track reconstruction efficiency is 96% over the complete momentum range

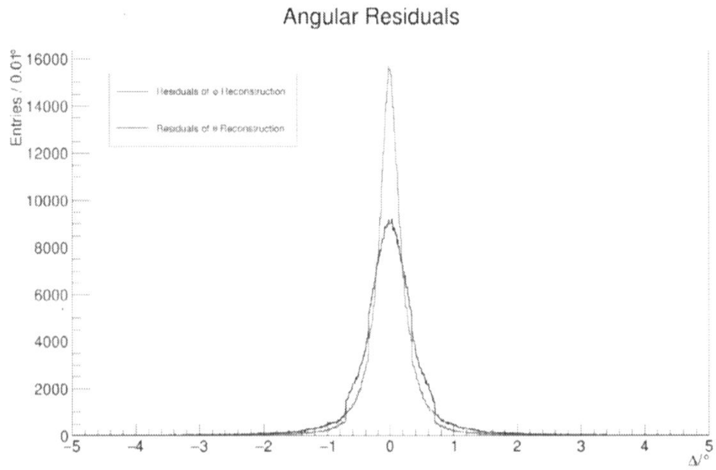

Figure 6. Angular residuals Δ of the reconstructed tracks, defined as difference between reconstructed and true angles, for φ (blue) and θ (red) of the cases. In 92% (φ) and 93% (θ), respectively, the reconstructed values show a deviation of less than 1° from the true track. The sharp edges edges at ±0.35° and ±0.7° in both distributions are caused by the discrete Hough space.

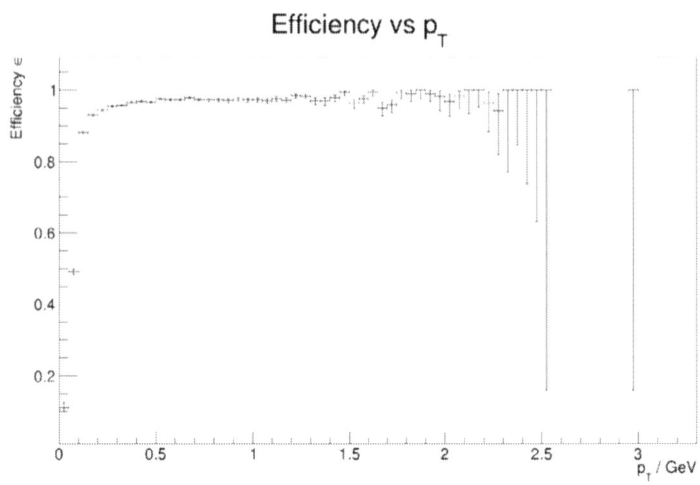

Figure 7. Track reconstruction efficiency as a function of the transverse momentum p_T. The sharp drop in the low-p_T region is caused by the fact that particles with $p_T < 80$ MeV do not reach the third layer of the SVD and thus do not produce the required minimum number of three SVD hits. The large error bars for $p_T > 1.5$ GeV are due to limited occurrence of particles with such a transverse momentum in the simulation. The average track reconstruction efficiency is 96% in the complete p_T-range shown.

expected for the decay products of the $\Upsilon(4S)$ resonance. The reduced efficiency for low-p_T tracks is primarily due to the fact that they only traverse one or two SVD layers and the DATCON algorithm requires at least three hits in different layers to identify a track to reduce combinatorics.

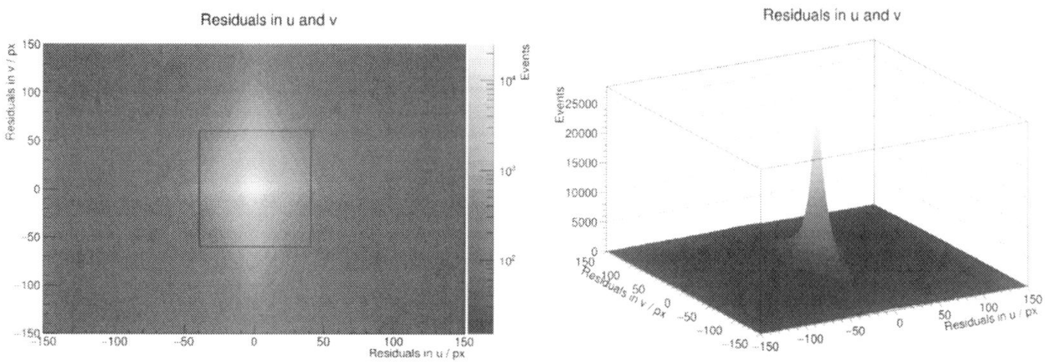

Figure 8. Residuals of the extrapolation to the PXD (measured in pixels) in local sensor coordinates u and v. The residual is defined as the difference between the local coordinates of the extrapolated hit and the local coordinates of the true simulated Monte-Carlo hit. The left side shows the residuals on a logarithmic scale and the ROI of 80×120 pixels around the MPH as a red box. The right side shows a 3D illustration of the residuals. With ROIs of size 80×120 pixels, the ROI finding efficiency is 94% with a data reduction factor of 15. To obtain a higher ROI finding efficiency, it is possible to further increase the ROI size while still reducing the data by a factor of about 10.

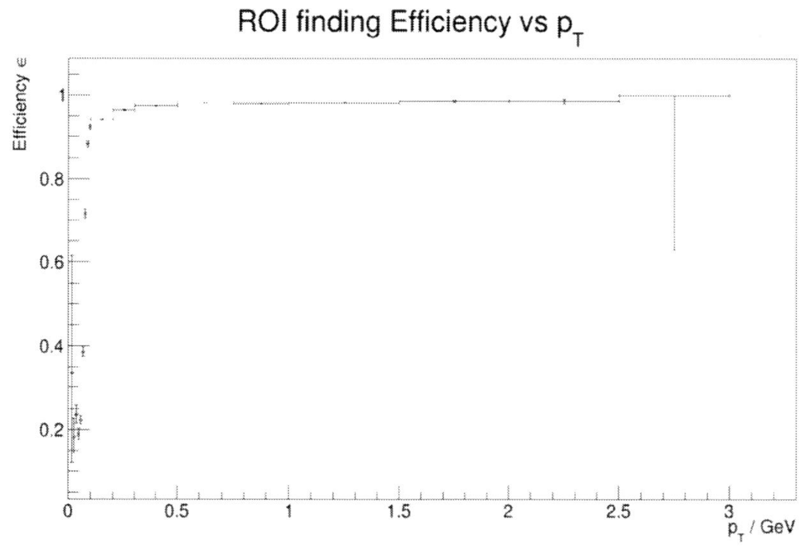

Figure 9. Efficiency of the ROI finding as a function of track p_T. Particles with a transverse momentum below 80 MeV cannot be found during tracking, and thus also no extrapolation is possible for these tracks. For tracks with $p_T > 100$ MeV, the ROI finding efficiency is above 90%, the average over the complete p_T-range is larger than 94%, with the data reduction factor being above 10.

Figure 8 shows the residuals of the extrapolated hits, defined as the difference between the local position of the extrapolated hit and the local position of the true MC hit, measured in pixels in the local coordinates u and v. Over 90% of the extrapolated hits are very close to the true hit position within

$\Delta R = \sqrt{u^2 + v^2}$ of 50 pixels, and 94% of all MPHs in the PXD are located such that the ROIs of size 80×1120 pixels ($u \times v$) calculated around the MPHs include the corresponding true PXD hits. As can be seen in Figure 8, the residual distribution is slightly wider in v-direction than in u-direction, which is why the ROI size is chosen to be larger in v-direction. Finally, Figure 9 shows the ROI-finding efficiency as a function of track p_T. The average ROI-finding efficiency over the whole p_T-range is larger than 94%, and for tracks with $p_T > 100$ MeV the algorithm is nearly 100% efficient. The median data reduction factor is about 15, leaving room to further optimise the ROI size and other aspects of the algorithm.

5 Conclusions and Outlook

In this article the performance of the DATCON system was presented: DATCON performs an FPGA-based online track reconstruction to define ROIs on the Belle II Pixel Detector using reconstructed hits from the Silicon Vertex Detector. This allows one to reduce the amount of data needed to be stored offline. The currently achieved median data reduction factor is 15. The track reconstruction efficiency that can be achieved with $\Upsilon(4S) \rightarrow B\bar{B}$ events is about 96% over the full p_T range, as determined by simulation. The efficiency of finding a true hit within a DATCON ROI is 94%. Future improvements on the performance could be obtained by tuning the ROI size using the p_T of the reconstructed tracks or by improving the clustering of the Hough space.

Acknowledgements

This work is supported by the German Federal Ministry of Education and Research (BMBF).

References

[1] R.H. Richter et al., Nucl. Instrum. Meth. **A511**, 250 (2003)
[2] T. Abe et al. (Belle-II) (2010), 1011.0352
[3] R. Itoh, T. Higuchi, M. Nakao, S.Y. Suzuki, S. Lee, IEEE Trans. Nucl. Sci. **60**, 3720 (2013)
[4] K. Lautenbach et al., JINST **12**, C06023 (2017)
[5] P.V.C. Hough, **C590914**, 554 (1959)
[6] D. Kim (Belle-II Software Group), Nucl. Part. Phys. Proc. **273-275**, 957 (2016)

Track reconstruction at LHC as a collaborative data challenge use case with RAMP

Sabrina **Amrouche**[1], Nils **Braun**[2], Paolo **Calafiura**[3], Steven **Farrell**[3], Jochen **Gemmler**[2], Cécile **Germain**[4], Vladimir Vava **Gligorov**[5], Tobias **Golling**[1], Heather **Gray**[3], Isabelle **Guyon**[6], Mikhail **Hushchyn**[7], Vincenzo **Innocente**[8], Balázs **Kégl**[9,10], Sara **Neuhaus**[11], David **Rousseau**[9], Andreas **Salzburger**[8], Andrei **Ustyuzhanin**[7,12], Jean-Roch **Vlimant**[13], Christian **Wessel**[14], and Yetkin **Yilmaz**[9,10]

[1] *University of Geneva, Switzerland*
[2] *Karlsruhe Institute of Technology, Germany*
[3] *Lawrence Berkeley National Laboratory, CA, USA*
[4] *LAL and LRI, Orsay, France*
[5] *LPNHE, Paris, France*
[6] *LRI and Université Paris-Saclay, France*
[7] *Yandex School of Data Analysis (YSDA), Moscow, Russia*
[8] *CERN, Geneva, Switzerland*
[9] *LAL, Univ. Paris-Sud, CNRS/IN2P3, Université Paris-Saclay, Orsay, France*
[10] *Center for Data Science, Université Paris-Saclay, France*
[11] *Max Planck Institute for Physics, Munich, Germany*
[12] *Higher School of Economics (HSE), Moscow, Russia*
[13] *California Institute of Technology, CA, Germany*
[14] *University of Bonn, Germany*

Abstract. Charged particle track reconstruction is a major component of data-processing in high-energy physics experiments such as those at the Large Hadron Collider (LHC), and is foreseen to become more and more challenging with higher collision rates. A simplified two-dimensional version of the track reconstruction problem is set up on a collaborative platform, RAMP, in order for the developers to prototype and test new ideas. A small-scale competition was held during the Connecting The Dots / Intelligent Trackers 2017 (CTDWIT 2017) workshop. Despite the short time scale, a number of different approaches have been developed and compared along a single score metric, which was kept generic enough to accommodate a summarized performance in terms of both efficiency and fake rates.

1 Data challenges and RAMP

Advances in machine-learning technology is moving towards a new paradigm in data processing design, in which the role of the individual will have to be redefined. In order to integrate the human creativity with the computational resources in an efficient way, various data-challenge platforms have been emerging, which provide a competitive and/or collaborative environment to crowd-source the analysis expertise.

Rapid Analytics and Model Prototyping (RAMP) [1] is an online data challenge platform, primarily motivated by the need to achieve a more systematic integration of the development. It aims to incorporate specific types of interactions between developers during the design process, and seeks insight into the evolution of the design. What distinguishes RAMP among various data challenge platforms is its focus on the progress rather than the result.

The RAMP includes many different types of data challenges, in addition to the track reconstruction challenge, to each of which any user can subscribe and submit several "solutions" for the formulated "problem". The problems can be of different types, such as classification, regression or clustering, in a variety of contexts such as chemistry, biology or climate research [1]. The distinguishing feature of RAMP is that the submission is the "code" to solve the problem, which runs remotely on a fully modularized workflow. For a given problem the users can implement methods for training and predicting, with any level of intermediate processes, as long as the methods return the expected format for the prediction. A screenshot of the submission sandbox is shown in Figure 1. The submissions are written in `python` and can use common libraries such as `numpy` [2], `pandas` [3], `scikit-learn` [4].

Figure 1. Screenshot of the RAMP sandbox window, with submission based on a basic DBSCAN clustering method [5]. Code submissions are made by either editing code on the page directly, or uploading the relevant files.

In order to test the code before submitting, the users are provided with a limited portion of the dataset that contains the relevant input and output columns, on which they can perform a quick cross-validation. The submitted code is then run in the backend to train and test on a larger dataset, and the results are displayed in an online table. Such modularization of the workflow allows the algorithms to be trained and tested on different datasets even after the data challenge, which will be the topic of the discussion in Section 6.

A typical hackathon with RAMP contains two phases: a closed phase and an open phase. In the closed phase, the participants cannot access each others submissions (although they may discuss among themselves), therefore each individual implements new code. In the open phase, participants can see and download each other's code, and make new submissions upon improvements on each other's ideas.

Taking into account the complexity of the track reconstruction problem and a short time period, the hackathon event during the CTDWIT 2017 workshop at LAL-Orsay started directly with the open phase in order to boost the first submission of the participants. This open phase also included a few test algorithms (without machine-learning techniques) as a guiding baseline.

2 Use case: Track reconstruction at the LHC experiments

Track reconstruction is one of the most important tasks in a high-energy physics experiment, as it provides high-precision position and momentum information of charged particles. Such information is crucial for a diverse variety of physics studies - from Standard Model tests to new particle searches - which require robust low-level information which can be further refined for a narrower and more specific physics context. Through the history of high-energy physics, there have been many different types of tracking detectors with very different design principles: from bubble chambers to time-projection chambers, from proportional counters to spark chambers, and more recently with silicon detectors... Although each of them have provided a different data topology, they all relied on the same principle: the small energy deposit of charged particles in well-defined locations, with particle trajectories being bent in an externally applied magnetic field. Future developments, like the High-Luminosity LHC [6, 7] which introduces a factor 10 increase in track multiplicity w.r.t. current LHC conditions, will make track finding very difficult.

A track reconstruction challenge [8] with a rather complete LHC detector simulation is being prepared by a number of co-authors and others. But for the limited time frame of this workshop, we propose a challenge where we reduce the track finding problem to two dimensions and emulate the topology of silicon detectors, such that there are few locations along the radius, and a very high precision along the azimuth. Such topology helps to simplify the track reconstruction problem to layer-by-layer azimuth determination, however, we hope to open room for further innovation as well.

3 Simulation

The challenge uses a simplified 2D detector model representative of an LHC detector with a toy model for particle generation and simulation, with sufficient complexity to make it non-trivial, with source code available in [9]. The detector consists of a plane with nine concentric circular layers with radii between 3.9 cm and 1 m, with a distribution typical of a full Silicon LHC detector. A homogeneous magnetic field of 2 Tesla perpendicular to the plane bends the charged particles so that a particle of momentum P (in MeV, with the convention $c = 1$) has a circular trajectory of radius $R = \frac{P}{0.3}$ (in millimeters). Layers have negligible thickness and each layer is segmented in azimuth with fine granularity. There are $2\pi R/pitch$ pixels in every layer, where $pitch$ is 0.025 mm for innermost layers 1-5 and 0.05 mm for outermost layers 6-9. Every "pixel" corresponds to the smallest detector unit defined by layer and azimuthal index.

A particle crossing a layer activates one pixel with a probability of 97%. The location of this activated pixel will be called a *hit* in the following. Since the detector has a very fine granularity in azimuth, the cases where two particles pass through a single pixel are rare (less than 0.2% probability). As a very simple simulation of hadronic interaction, a particle has a 1% probability to stop at each layer. Given all this, participants cannot assume a particle will leave a hit in each layer.

A charged particle crossing matter will accumulate many small random deviations due to interaction with atomic nuclei. This multiple scattering phenomenon is simulated by attributing to each layer a radiation length x_0 of 2%, such that a particle crossing a layer is deviated by an angle drawn from a gaussian centered at zero and of width $\sigma_{MS} = 13.6\sqrt{x_0}/P$ in radians. Thus trajectories deviate from perfect circular arcs. The energy loss is neglected, so that particles keep the same absolute momentum, hence the same local curvature, throughout the trajectory.

For each event, a Poisson distribution is sampled to determine the number of particles, with an average of 10 particles per event (3 orders of magnitude less than what is expected at HL-LHC). The particles are sampled uniformly in azimuth and momentum, with bounds on the momentum between 300 MeV and 10000 MeV. The lowest momentum is such that particles cannot loop in the detector. Each particle originates from a different vertex that is also randomly sampled from a normal distribution around the origin of width 0.667 mm, as a simple simulation of particles coming from B-hadrons or τ-leptons, so that particles cannot be assumed to come exactly from the origin. No additional random background hits have been simulated.

The data provided to the participant is one csv file with one line per hit corresponding to 10000 events, while a separate 10000 event file is kept private for the evaluation of the submissions by the RAMP platform. Hits are grouped per event in a random order. The following features are provided:

event number : arbitrary index, unique identifier of the event

particle number : arbitrary index, unique identifier of the ground truth particle having caused that hit. This number is hidden from the prediction method.

hit identifier : index of the hit in the event (implicit variable)

layer number : index of the layer where the hit occurs, from 1, the innermost layer, to 9, the outermost layer

phi number index of the pixel corresponding to the hit, counting clockwise, from 0, at $\phi = 0$ to the maximum, which depend of the layer

x : horizontal coordinate of the hit

y : vertical coordinate of the hit

It should be noted that although one can easily translate layer and phi information into x and y (and vice versa), both are provided in order to help participants to gain time. The user code should return for each event of the test dataset a track number for each hit, the track number being an arbitrary unique (in the event) number corresponding to a cluster of hits.

4 Scoring

The programs of LHC experiments span a diverse range of physics topics, each with a different selection and use of data. Although tracks are used in almost all analysis applications, they are employed in very different ways: sometimes as an important part of the signal, sometimes as an indicator of background. The overall goal of track reconstruction is to achieve the best correspondence between particles and tracks, where the term "particle" refers to the true particles and "track" refers to the information deduced from the detector by the reconstruction algorithm. This correspondence is never perfect: Sometimes a particle is missed by the algorithm and this is referred to a lack of *efficiency*, and sometimes a track is reconstructed out of a coincidental alignment of hits without corresponding to a genuine particle, which is referred to as *fake*.

Depending on the analysis goals, one may be focused on very different aspects of the reconstruction: some analyses need minimum combinatorial fake rate, some need the highest efficiency or the best momentum resolution. Ultimately, the figure of merit is defined by the systematic uncertainty on the measured quantity which, as stated, varies for every study. It is therefore difficult to define in advance the perfect track reconstruction score that fits all.

Ultimately the score that was used in the hackathon was based on "fraction of hits that are correctly matched to a particle in a given event", and more precisely defined as follows:

1. Loop over particles to associate each particle to a set of tracks
2. Choose the track with the most hits in the particle, and associate it to the particle. The result will be many(track)-to-one(particle).
3. Loop over tracks and check which tracks are matched to the same particle.
4. In the case of multiple matching, remove all tracks (and their corresponding hits) from the list except the one with the most hits in the particle.
5. Count the number of remaining hits (that are all associated to tracks by construction), and divide it by the initial number of hits.
6. Assign the score of the event with the above, and average it over all the events.

The code for the score function is provided to the participants as a part of the starting kit. Figure 2 illustrates the scoring procedure on a single event. Every event is scored separately, and the final result is averaged over events (all events having equal weight, regardless of number of hits).

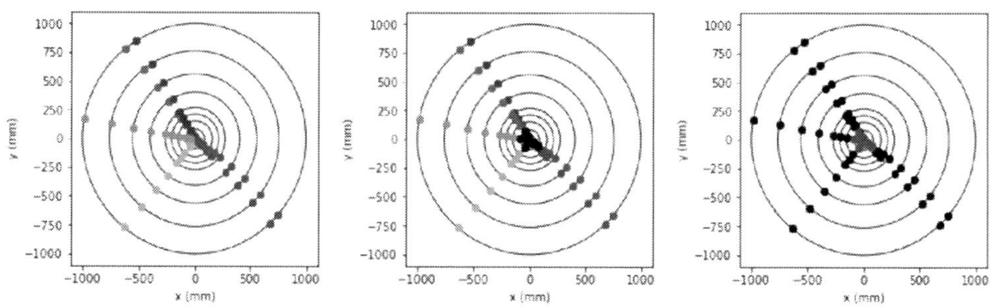

Figure 2. Illustration of the scoring procedure. First plot shows the ground truth and the middle plot is the reconstructed output, after indices of tracks (represented by colors) are matched to the input. The black dots in the last plot shows the hits that are considered to be correctly assigned. For this particular event, the score is 0.81.

It is important to remark that, it is possible to think of analyses tasks where the trade-off between an optimum fake rate and efficiency can be contrary to the relative contribution of these properties on the clustering score. The correspondence between the RAMP score and the actual track reconstruction efficiency and fake rate is studied for various algorithms and various samples. Here the efficiency is defined as the number of well-reconstructed tracks divided by the number of true particles, where well-reconstructed means that at least 70% of the hits belonging to the same particle are clustered successfully. Similarly, fake rate is defined as the number of fake tracks divided by the number of true particles, when a fake track is defined as a cluster of at least 5 hits in which no more than 70% of the hits belong to the same true particle.

In Figure 3, several submissions for a specific algorithm are plotted, in which the participant only knows the score and tries to optimize that in every update. According to this study, an improvement in the score represents a monotonic improvement on both the efficiency and fake rate, therefore the score is reliable when upgrading the algorithms to achieve both better efficiency and less fakes. The robustness of the score function, in terms of efficiency and fake rate optimization in various samples, is further discussed in Section 6.

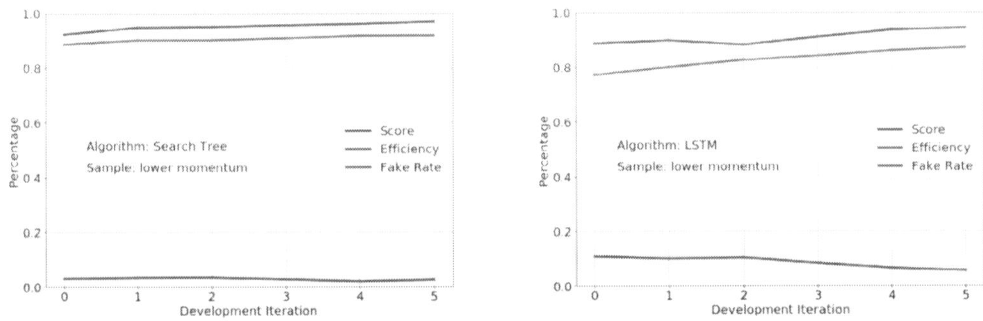

Figure 3. Illustration of the score as a performance indicator during the development process. Each panel corresponds to the submissions with a specific approach, incrementally being improved.

5 Example submissions

There were many interesting submissions during the hackathon, of which some with the highest scores are displayed in the leaderboard shown in Figure 4. At the end of the first day, the scores already started to approach towards a saturation around 97% as shown in Figure 5. Some of the approaches were totally independent, although some were a combination of existing baseline examples with improved features. We will now illustrate some of the example submissions which displayed interesting lines of development. The source code for these submissions can be accessed by signing up on the RAMP [1].

5.1 Untrained baseline examples

We started the hackathon with three baseline examples which we kept as simple as possible. These algorithms were not data-driven, therefore the training function was left blank.

Among these three, two algorithms were known to be only valid for the assumption that the circular tracks have very small curvature and therefore can be approximated by straight lines. They all assumed the vertex position to be at the exact origin. In the `linear approximation` algorithm, the compatibility of hits with a straight line is tested and hits are clustered accordingly. In the `nearest neighbors` algorithm, the hits at each next layer are matched to the closest hit in the previous layer.

The third algorithm was a simplified version of Hough transform, in which the individual hit positions were projected on to the track parameter space of $P \times \phi$. The actual implementation had the further simplification that the binning and searching in the $P \times \phi$ space was restricted to bands around hit pairs in the first two layers, therefore it can be considered as a "seeded" Hough transform approach.

DOI: 10.1051/epjconf/201715000015

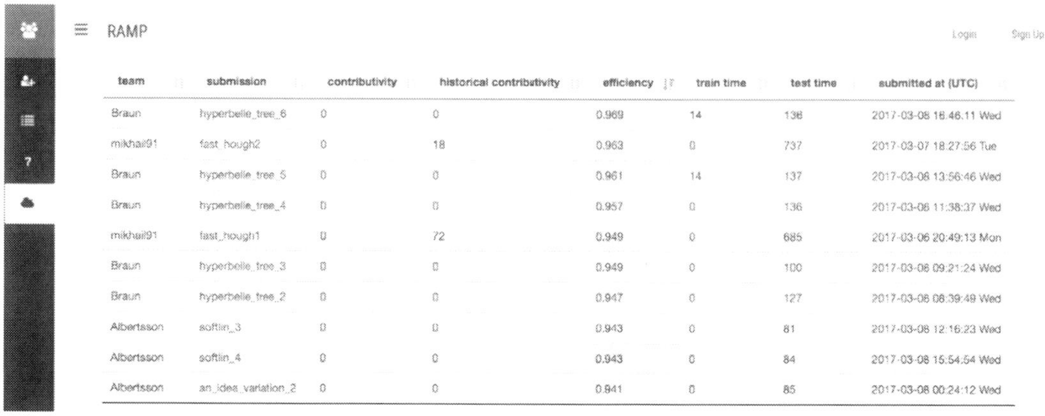

RAMP

team	submission	contributivity	historical contributivity	efficiency	train time	test time	submitted at (UTC)
Braun	hyperbelle_tree_6	0	0	0.969	14	138	2017-03-08 16:46:11 Wed
mikhail91	fast_hough2	0	18	0.963	0	737	2017-03-07 18:27:56 Tue
Braun	hyperbelle_tree_5	0	0	0.961	14	137	2017-03-08 13:56:46 Wed
Braun	hyperbelle_tree_4	0	0	0.957	0	136	2017-03-08 11:38:37 Wed
mikhail91	fast_hough1	0	72	0.949	0	685	2017-03-06 20:49:13 Mon
Braun	hyperbelle_tree_3	0	0	0.949	0	100	2017-03-08 09:21:24 Wed
Braun	hyperbelle_tree_2	0	0	0.947	0	127	2017-03-08 08:39:49 Wed
Albertsson	softlin_3	0	0	0.943	0	81	2017-03-08 12:16:23 Wed
Albertsson	softlin_4	0	0	0.943	0	84	2017-03-08 15:54:54 Wed
Albertsson	an_idea_variation_2	0	0	0.941	0	85	2017-03-08 00:24:12 Wed

Figure 4. Screenshot of the RAMP leaderboard, showing top 10 submissions as of March 9, 2017.

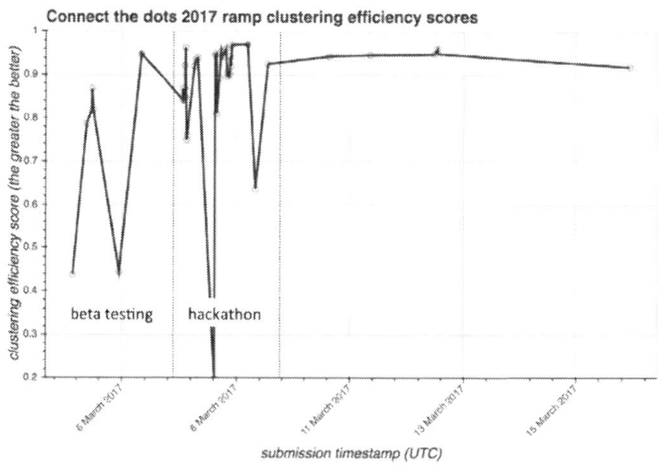

Figure 5. Development of algorithms during the hackathon. Although the submissions from the beta-testing period already had a high score, the participants were able to implement faster algorithms with even better scores.

Although the score of the Hough transform algorithm was one of the highest (0.97), it is slow (0.16s/event). The algorithms described below are faster, with similar scores.

5.2 The $\Delta\phi$ algorithm

The **`delphi_multilayer`** submission is an evolution of the baseline example, to which it owes its simplicity and good CPU performance. It is a $O(N^2)$ algorithm which builds a track by aligning the direction vector of hit pairs from consecutive layers. In the (R,ϕ) plane this means cutting on the "second derivative" $C_n = \phi_n - 2\phi_{n-1} + \phi_{n-2}$, where ϕ_n is the azimuth of the hit in the n-th layer. Five simple improvements allowed to increase the algorithm efficiency from 84% to 96% with small CPU performance degradation:

1. Allow tracks to have missing layers, including in the innermost and outermost layers;
2. Search for tracks bidirectionally: first outside-in, then inside-out;
3. Introduce a layer-dependent cut on direction vector alignment to take into account sensor pitch, and multiple scattering;
4. Follow the curvature of the track when comparing vector alignments, cutting on the difference between the current C_n and the running average $< C >$;
5. Allow hits to be shared among tracks, with a "sharing cost" tuned to limit fake tracks.

5.3 Inward search tree

The winning contribution (submitted as `hyperbelle_tree`) in the hackathon was inspired by a typical track reconstruction algorithm used in high energy physics, e.g. in the software currently developed for the Belle II experiment [10]. The implemented code is similar to a combinatoric Kalman filter and loosely based on the principles of a Monte Carlo Search Tree. To keep the implementation simple, the update step of the Kalman filter which, in real-world applications defines how well the algorithm works, was not implemented.

The algorithm starts with a random hit (the seed) on the outermost layer and builds up a tree using the hits on the next layer. A weight is calculated for each hit, which will be discussed below. Using only the hit with the maximal weight and those, which have a weight close to the maximal one, a new track candidate is built and the procedure is repeated until the innermost layer is processed.

The weight calculation is the crucial part in this algorithm. In this example implementation, the azimuthal angle difference between two successive hits is sampled on training data, where the truth information is known. From this histogrammed data, a probability for a given phi difference is extracted and used as a weight.

The tree search may produce multiple track candidates starting with the same seed hit, so only the candidate with the most consistency to a circle is stored. As there are no background hits, the tree search is continued until all hits of the event are assigned. To cope with hit inefficiencies, tracks which do not include hits of all layers are fitted together pairwise and tested for their fit quality. Good combinations are merged together.

As one part of the challenge was also to write fast algorithms, a runtime optimization of the implementation was performed. Caching for the hits on each layer was implemented and heavy calculations and loops were realized using `numpy` functionality.

5.4 LSTM based algorithm

The LSTM submission [11] is based on a Long Short Term Memory neural network algorithm [12]. LSTMs can be powerful state estimators due to their effective usage of memory and non-linear modeling of sequence data. For this problem, the detector layers form the sequence of the model, and the predictions are classification scores of the hits for a given track candidate.

The input for the LSTM model was prepared as follows. First, the hit data is histogrammed for each detector layer in the ϕ coordinate with 200 binary bins. A bin has a value of 1 if it contains a hit, otherwise 0. The data is thus a sequence of 9 vectors of length 200. The model operates on one track candidate at a time, so track "seeds" were identified using only the hits in the first detector layer. Each training sample input is prepared by using only the seed hit in the first layer plus all of the event's hits in the remaining layers. The corresponding training sample target contains only the hits of the true track.

For each training sample, the LSTM model reads the input sequence and returns a prediction sequence for each corresponding detector layer. Each vector of the output sequence is fed through a fully connected neural network layer and a softmax activation function to give a set of "probability" scores over the ϕ bins on that detector layer. For the bins that have hits in them, this score quantifies confidence that a hit belongs to the track candidate. The training objective, or cost function, is a categorical cross-entropy function.

Once trained, the model's predictions can be used to assign labels to hits. Each event will contain a number of track candidates and thus each hit will have an associated score from the model predictions on each of those candidates. The assignment is made by choosing the candidate that gave the highest prediction score to the hit.

6 Tests with post-hackathon samples

As mentioned earlier, two of the baseline algorithms were only valid under the conditions of high average momentum and low detector occupancy. These algorithms are expected to fail if we test on a sample that moves out of these assumptions.

Some of the other algorithms, however, can learn the patterns and adapt to the conditions, if we also train on a sample that was prepared differently. This is how RAMP is set up, and here we illustrate the advantage of using the platform for the full analysis workflow rather than submitting a fixed solution.

In addition to the sample used in the hackathon (described in Section 3), we produced two alternative samples which are more challenging in terms of the physics: lower momentum (more curved) tracks, and higher multiplicity (hence detector occupancy), summarized in Table 1.

Table 1. Parameters of the input samples.

sample	multiplicity	momentum range (MeV/c)
hackathon	10	[300,10000]
high-mult	20	[300,10000]
low-mom	10	[100,1000]

Since the linear approximation algorithm relies on the assumption that the tracks are almost straight lines, the algorithm starts to fail on a sample of tracks with more curvature. As shown in Figure 6, the test with high-multiplicity sample performs similarly to the hackathon sample, whereas the low-momentum sample displays a big degradation in the performance.

The Hough transform takes into account the curvature of the tracks, however it also displays (Figure 7) a slight degradation of performance in the high-multiplicity and low-momentum samples.

Performances of the $\Delta\phi$ algorithm (Figure 8) degrades slightly when going from the hackathon to the high-multiplicity sample, and significantly on the low-momentum sample due to difficulties accomodating the larger curvature of the tracks.

The Inward search tree displays (Figure 9) similar performance in the hackathon and high-multiplicity samples, and a slight decrease in the score in the low-momentum sample. The efficiency degrades less than the score, which can be interpreted as being able to reconstruct the same tracks missing some of the hits along the track.

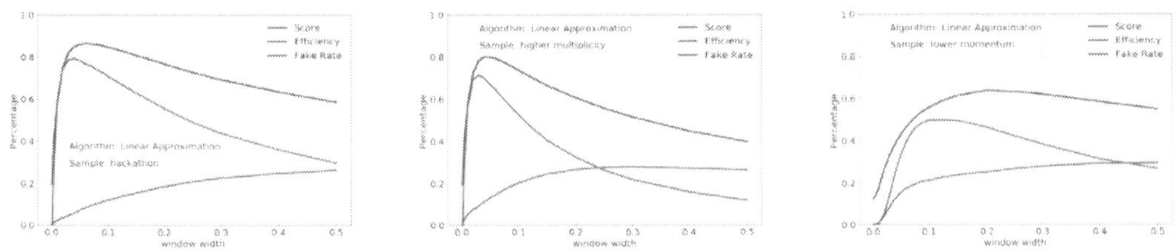

Figure 6. Performance of the linear approximation algorithm as a function of the angular window width, which is the only parameter of the algorithm. The left figure displays the performance in the nominal hackathon sample, the center and right figures display the performances in the modified "high-multiplicity" and "low-momentum" samples, respectively. The optimum window size highly depends on the momentum distribution.

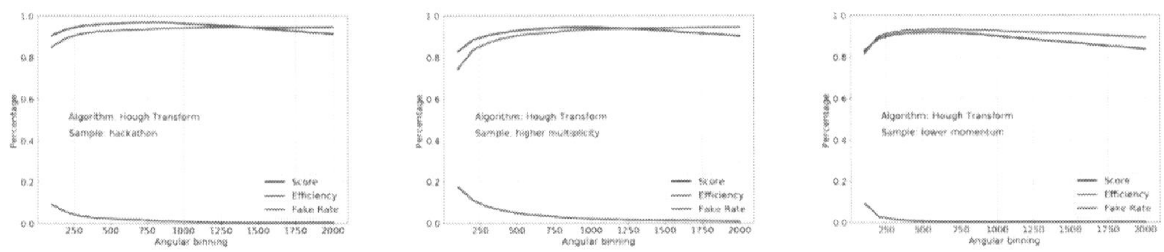

Figure 7. Performance of the Hough transform algorithm as a function of the angular binning. The left figure displays the performance in the nominal hackathon sample, the center and right figures display the performances in the modified "high-multiplicity" and "low-momentum" samples, respectively. Although finer binning appears to almost monotonically increasing the efficiency, it may cause the loss of hits on the tracks, hence the score has slightly steeper dependence.

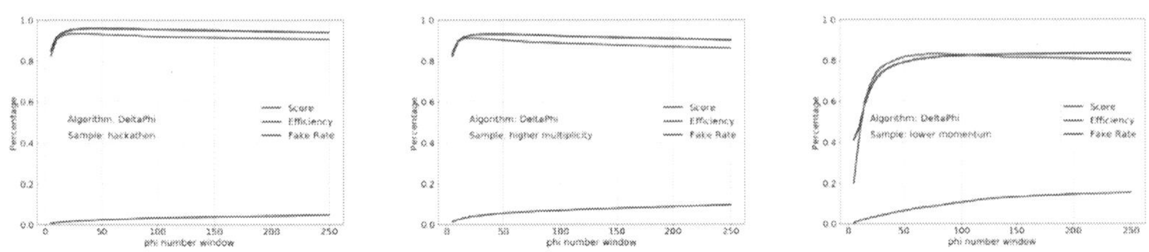

Figure 8. Performance of the $\Delta\phi$ algorithm as a function of the angular window width, which is the only parameter of the algorithm. The left figure displays the performance in the nominal hackathon sample, the center and right figures display the performances in the modified "high-multiplicity" and "low-momentum" samples, respectively. The optimum window size highly depends on the momentum distribution.

The LSTM algorithm displays (Figure 10) a small degradation of efficiency in the high-multiplicity sample. The low-momentum sample is out of validity of the submitted implementation. Due to high curvature, this sample has several tracks not having hits in all layers or two hits in one layer, and the algorithm can be improved in the future to incorporate such cases in a better way.

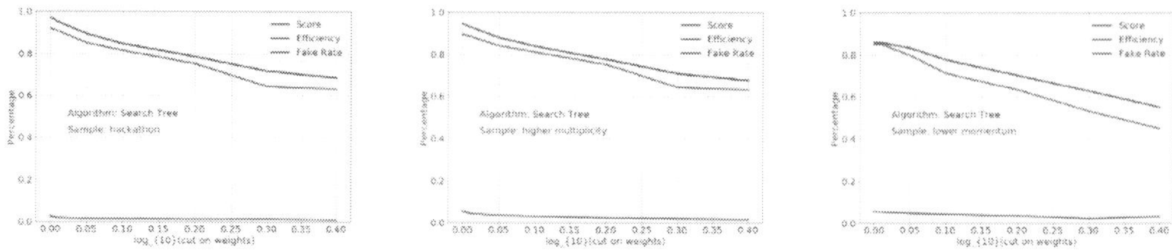

Figure 9. Performance of the inward search tree algorithm, as a function of the exponent of the cut applied on the weights. The left figure displays the performance in the nominal hackathon sample, the center and right figures display the performances in the modified "high-multiplicity" and "low-momentum" samples, respectively.

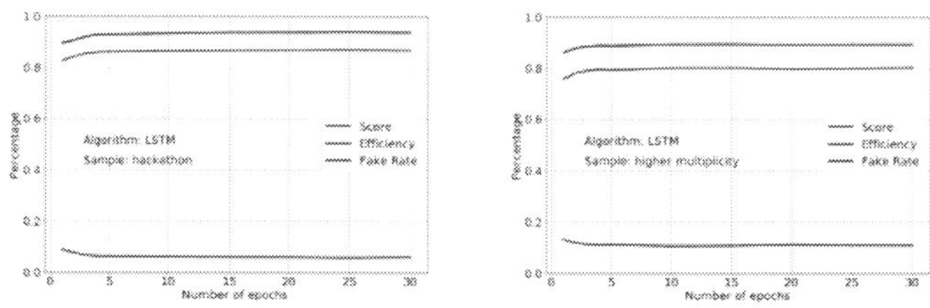

Figure 10. Performance of the LSTM algorithm as a function of number of epochs used in the training. The left figure displays the performance in the nominal hackathon sample and the right figure displays the performance in the modified "high-multiplicity" sample.

7 Conclusions and outlook

The problems for the CTDWIT 2017 hackathon were designed in a simplified way in 2D in order to allow the users to experiment on novel approaches, instead of being confined to regular methods of track reconstruction, within a 36 hours hackathon. In particular, the 2-dimensional setup simplified the visualisation of both the problem and solution, and the high average momentum and low multiplicity setup of the input samples made it easier to find a starting point for the development. With about 30 submissions in two days, the hackathon established a proof of principle for a wider scale data challenge for a more realistic experimental setup.

The post-hackathon test of the analysis in much more challenging physics conditions illustrated that it is possible to think ahead of the available data and implement robust solutions even if some features of the data are unknown. As expected, the algorithms that relied on the assumptions limited to the provided data failed, however a number of them maintained a high score. The experimental aspect of the hackathon was the observation of the contrasts of these approaches and their relative performances.

Acknowledgement

This project has received funding from the European Union's Horizon 2020 research and innovation programme under grant agreement No 654168

References

[1] *Rapid analytics and model prototyping* (2017), http://ramp.studio/

[2] S. van der Walt, S.C. Colbert, G. Varoquaux, *The numpy array: A structure for efficient numerical computation, computing in science & engineering* (2011), http://www.numpy.org/

[3] W. McKinney, *pandas : powerful python data analysis toolkit* (2011), http://pandas.sourceforge.net/

[4] F. Pedregosa, G. Varoquaux, A. Gramfort, V. Michel, B. Thirion, O. Grisel, M. Blondel, P. Prettenhofer, R. Weiss, V. Dubourg et al., Journal of Machine Learning Research **12**, 2825 (2011)

[5] M. Ester, H.P. Kriegel, J. Sander, X. Xu, *A density-based algorithm for discovering clusters in large spatial databases with noise* (AAAI Press, 1996), pp. 226–231

[6] The ATLAS Collaboration,CERN-LHCC-2015-020, *"ATLAS Phase-II Upgrade Scoping Document"* (2015)

[7] The CMS Collaboration, CERN-LHCC-2015-019, *"CMS Phase-II Upgrade Scoping Document"* (2015)

[8] *The Tracking Machine Learning Challenge* (2017), https://github.com/trackml

[9] *TrackMLRamp Simulation* (2017), https://github.com/ramp-kits/HEP_tracking/tree/ramp_ctd_2017/simulation

[10] T. Abe et al. (Belle-II) (2010), arXiv:1011.0352

[11] S. Farrell, *The HEP.TrkX Project: deep neural networks for HL-LHC online and offline tracking*, in *Connecting The Dots / Intelligent Trackers* (2017)

[12] S. Hochreiter, J. Schmidhuber, Neural Comput. **9**, 1735 (1997)

FPGA-Based Tracklet Approach to Level-1 Track Finding at CMS for the HL-LHC

Edward **Bartz**[2], Jorge **Chaves**[1], Yuri **Gershtein**[2], Eva **Halkiadakis**[2], Michael **Hildreth**[3], Savvas **Kyriacou**[2], Kevin **Lannon**[3], Anthony **Lefeld**[4], Anders **Ryd**[1], Louise **Skinnari**[1], Robert **Stone**[2], Charles **Strohman**[1], Zhengcheng **Tao**[1], Brian **Winer**[4], Peter **Wittich**[1], and Margaret **Zientek**[1,a]

[1]*Cornell University, Ithaca, NY, USA*
[2]*Rutgers University, New Brunswick, NJ, USA*
[3]*University of Notre Dame, South Bend, IN, USA*
[4]*Ohio State University, Columbus, OH, USA*

Abstract. During the High Luminosity LHC, the CMS detector will need charged particle tracking at the hardware trigger level to maintain a manageable trigger rate and achieve its physics goals. The tracklet approach is a track-finding algorithm based on a road-search algorithm that has been implemented on commercially available FPGA technology. The tracklet algorithm has achieved high performance in track-finding and completes tracking within 3.4 μs on a Xilinx Virtex-7 FPGA. An overview of the algorithm and its implementation on an FPGA is given, results are shown from a demonstrator test stand and system performance studies are presented.

1 Introduction

CERN's LHC accelerator complex will undergo major upgrades, planned for 2025, to increase the instantaneous luminosity up to around 7.5×10^{34} cm^{-2}s^{-1} [1]. This era of the High Luminosity LHC (HL-LHC) will yield an average number of overlapping proton-proton collisions per bunch crossing (pileup or PU) between 140 and 200. The LHC collides proton bunches every 25 ns. If every collision were kept, this would mean roughly 20 to 40 Tbps would need to be stored. However, only a small fraction of these collisions are of interest for further study. The CMS [2] physics program for the HL-LHC requires being able to intelligently select or trigger on collisions that could contain interesting physics scenarios. The trigger system reduces the 40 MHz collision rate to a manageable data storage rate of 400 Hz.

An integral part of reducing the rate is the hardware Level-1 (L1) trigger system. At the HL-LHC, the L1 trigger rates for objects such as single muons, electrons, or jets will exceed the current front-end capabilities. Increasing the transverse momentum (p_T) trigger thresholds for these objects limits the physics potential of the HL-LHC and no longer sufficiently reduces the data rate. Integrating charged particle tracking from the silicon tracker into the L1 trigger will improve lepton identification and momentum measurements and provide track isolation and vertex identification. These additional handles have the potential to reduce the L1 rates while maintaining reasonable trigger thresholds. One

[a]e-mail: mez34@cornell.edu

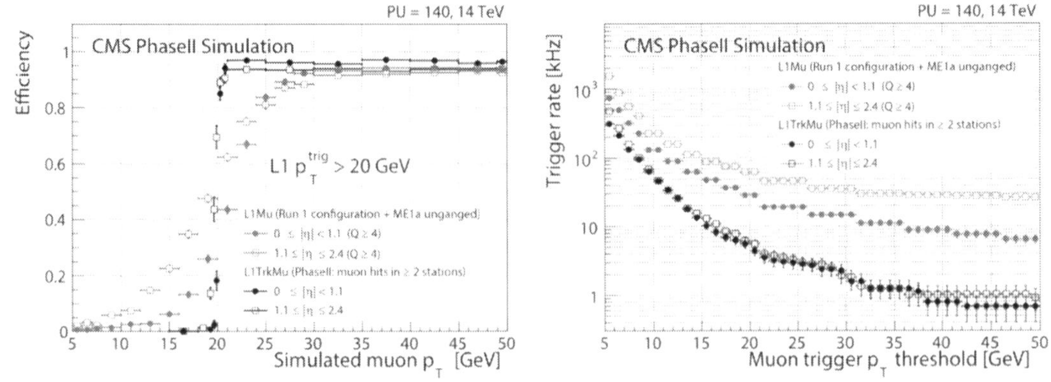

Figure 1. The efficiency of a single muon trigger with $p_T > 20$ GeV as a function of p_T (left) and the trigger rate as a function of the muon trigger threshold (right), shown for the stand-alone muon trigger (red) and when including L1 tracking from the silicon tracker (black) for various η ranges [3].

example of this is shown in Figure 1, where efficiency (left) and rate (right) are plotted for a single muon trigger. The red curves show the behavior of a stand-alone L1 muon trigger system, while the black curves show the performance of the same triggers when including L1 tracking. In particular, adding L1 tracking improves the momentum measurement which translates to a sharp turn-on curve at the trigger threshold and a reduced trigger rate [3]. The space available to store each event in the buffer limits the time allowed for the trigger decision. The L1 trigger decision must occur within 12 μs. In order to leave time for correlating tracks with other physics objects and the trigger decision, the track reconstruction must be made in approximately 4 μs.

The upgraded ("Phase 2") CMS detector will include an all silicon tracker that is designed to integrate tracking into the L1 trigger decision. The new tracker consists of a pixel detector, which is not used in L1 tracking, and an outer tracker that has a central barrel and two endcaps. The outer tracker will have the unique feature of "p_T modules" that provide p_T discrimination at the level of the front-end readout electronics. In the detector's magnetic field, a charged particle's bend (and consequent hit pattern) depends on its p_T. Correlating hits between two closely spaced sensors thus provides a coarse p_T measurement. A p_T module consists of two sensors, either instrumented with pixels or strips, that are mounted 1–5 mm apart. Two different modules are used: pixel-strip (PS) modules in the inner layers of the barrel and inner half of the endcaps, and strip-strip (2S) modules in the outer layers of the barrel and outer half of the endcaps. Further details on the layout and module design can be found in Ref. [3]. The pixelated sensor of the PS modules provides a precise measurement of the displacement along the beam line (z-axis), which enables primary vertex reconstruction at the L1 level. Correlated hits ("stubs") in p_T modules are required to be consistent with a $p_T > 2$ GeV track originating from the interaction point. Since most minimum bias events have low p_T tracks, this provides a factor of $\approx 10\%$ data reduction already. The stubs are then sent to the L1 track finding system.

To summarize, the goals of the L1 tracking system are to reconstruct the trajectories of charged particles that have $p_T > 2$ GeV and identify the z_0 of the track (z coordinate where the track intercepts the z-axis) with about 1 mm precision, similar to the average separation of vertices in events with an average PU = 140. Additionally, the L1 track finding must be completed within 4 μs. There are three possible CMS implementations to L1 tracking: (i) the "tracklet" approach, a road-search algorithm implemented using commercially available FPGAs and the approach that is presented here, (ii) a Hough-transform based approach also using FPGAs [4], and (iii) an associative memory based approach using a custom ASIC [5]. The next sections will detail the tracklet algorithm, implementation on hardware, performance results, and some projections towards making a full system for completing L1 tracking at CMS at the HL-LHC.

2 Tracklet Algorithm

A sketch of the algorithm steps, which are detailed below, is shown in Figure 2. The tracklet approach starts by forming track seeds (tracklets) from pairs of stubs in adjacent barrel layers or endcap disks. The tracklet is an initial estimate of the tracklet parameters calculated from these two stubs using the interaction point as a constraint. A candidate tracklet must be consistent with a $p_T > 2$ GeV track that originates within $|z_0| < 15$ cm. The seeding is performed for several combinations to provide good coverage of the entire pseudorapidity (η) range of the detector. The tracking efficiency for different seeding combinations is shown for single muons in Figure 3, using an integer-based C++ emulation of the algorithm as it would be implemented on an FPGA. In the current implementation of the algorithm, seeding includes pairs between layers 1+2, 3+4, and 5+6, and between disks 1+2 and 3+4.

The track parameters of the tracklets are then projected to other layers and disks to search for consistent stubs. When the tracklets are projected to other layers/disks, the search for matching stubs occurs in predetermined search windows, derived from residuals between projected tracklets and stubs. The projection of the tracklets occurs both inwards and outwards (i.e. to and from the interaction point). If a stub is found that is consistent with the original tracklet's parameters, the matched stub is included in the track candidate and the difference between the projected tracklet position and the matched stub position is stored.

A linearized χ^2 fit is performed using all stubs in the track candidate - the stubs used to make the original tracklet plus the matched stubs. The track fit uses pre-calculated derivatives and the projection-stub differences. The linearized χ^2 fit corrects the initial tracklet parameters giving the final track parameters: p_T, η, z_0, the azimuthal angle at the closest approach (ϕ_0), and optionally the impact parameter of the transverse plane (d_0). Because seeding is performed for multiple seeding combinations, a single track may be found several times. Duplicated tracks are removed by comparing the found tracks in pairs, comparing the number of independent and shared stubs.

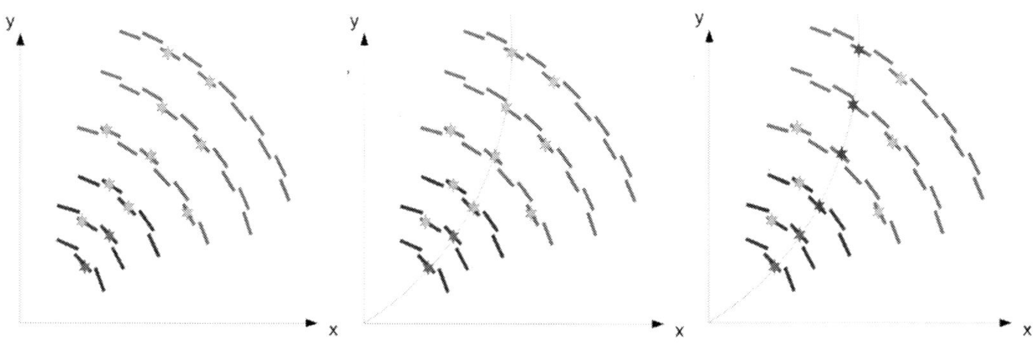

Figure 2. In the first step (left) of the algorithm a pair of stubs (red) in adjacent layers are combined to form a tracklet. The trajectory of the tracklet is projected (middle) to the other layers. Stubs in the other layers that are close to the projection (green) are selected as matches (right) to the tracklet. Final track parameters are calculated using all associated stubs.

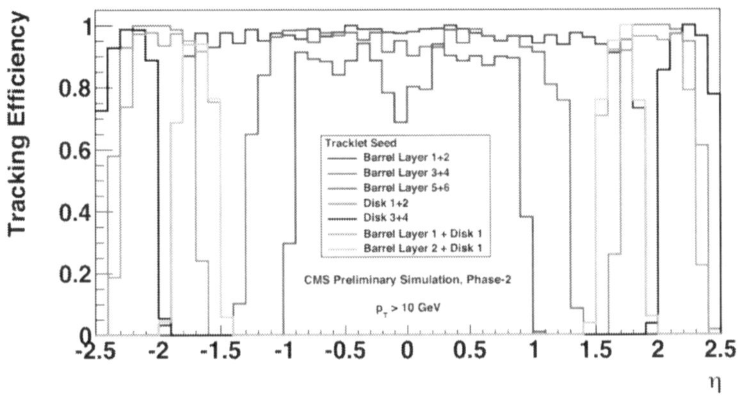

Figure 3. Efficiency for finding a single muon track for several different seeding layer/disk combinations as a function of η. Here it is seen that the combination of the different seeding pairs provides coverage of the full η range of the detector [6].

3 Hardware System

To address the challenging amount of data and limited processing time available, the tracklet hardware configuration relies on massively parallelizing the data processing. The main parallelization is the division of the detector into sectors in the r-ϕ plane. The current project uses 28 ϕ sectors. This was chosen so that tracks with $p_T > 2$ GeV span a maximum of two sectors. This limits the need for data transfer between sectors to the nearest neighboring sector on each side. Tracklets that project to a neighboring sector are sent there for tracklet-stub matching. A dedicated processing board is used for each ϕ sector. A small amount of data is duplicated every other layer to avoid gaps in the track-finding.

To allow for more time for data processing, the whole 28 ϕ sector system is replicated n times using a round robin time multiplexing approach. Each independent time multiplexed system receives a new event every $n \times 25$ ns. The choice of n is driven by a balance of cost, efficiency and needed processing power. For the full system, $n = 4 - 8$ are considered reasonable choices. The current implementation assumes a time multiplexing factor of $n = 6$, so a new event is received every 150 ns.

3.1 Hardware Implementation

The tracklet algorithm is implemented in the firmware as several processing steps (names in bold) [7]:

- *Stub organization*: (1) Sort input stubs by layer (**Layer Router**), and (2) into smaller units in z and ϕ called "virtual modules" (**VM Router**).
- *Tracklet formation*: (3) Select possible pairs of stubs (**Tracklet Engine**), and (4) calculate the tracklet parameters (**Tracklet Calculator**).
- *Projection organization*: (5) Transmit projections that point to neighboring sectors (**Projection Transceiver**), and (6) route the projections based on their virtual module (**Projection Router**).
- *Stub matching*: (7) Match projections to stubs (**Match Engine**), and (8) calculate the tracklet-stub difference in positions (**Match Calculator**), and (9) transmit matches back to original sector (**Match Transceiver**).
- *Track fit*: (10) Fit final tracks (**Track Fit**), and (11) remove duplicate tracks (**Duplicate Removal**).

The largest combinatoric challenges occur at tracklet formation and match finding. With an average PU = 140, there are ≈ 60 stubs per layer per ϕ sector, this would yield ≈ 3600 candidate tracklets per seeding combination [7]. By dividing each ϕ sector into smaller virtual modules, the tracklet formation and match finding processes are further parallelized. Furthermore, only a small fraction of virtual module pairs are consistent with $p_T > 2$ GeV and $|z_0| < 15$ cm tracks. This reduces the number of stub combinations that need to be tried at each of these steps. Consequently, the number of tracklets per ϕ sector is reduced to ≈ 20 per seeding combination.

All of the processing steps above read from memories filled by the previous step and write the output to another set of memories. Currently all processing steps are synchronized to a single common 240 MHz clock. By construction, the system is fully pipelined and operates at a fixed latency. When a new event arrives (currently every 150 ns) the previous event moves to the next processing step. This necessarily implies that a given step can only perform a fixed number of operations. If the time limit is reached, processing on the remainder of the data for that step stops, meaning any remaining data must be truncated. The effect of truncation on the system is minimal, and more details on the performance with truncation are presented later.

3.2 Hardware Demonstrator

The full tracklet algorithm, including all processing and transmission steps, has been implemented in firmware. Two complete implementations - one for half the barrel (+z) and one for a quarter of the barrel plus the forward endcaps - were used to demonstrate the feasibility of this approach for the full η range of the detector. A sketch of the upgraded CMS tracker region covered by the each of the implementations is shown in Figure 4. Note, the layout of the upgraded tracker will actually have tilted modules in the inner barrel layers with $\eta > 0.6$. The porting of the tracklet algorithm to the updated geometry is almost complete.

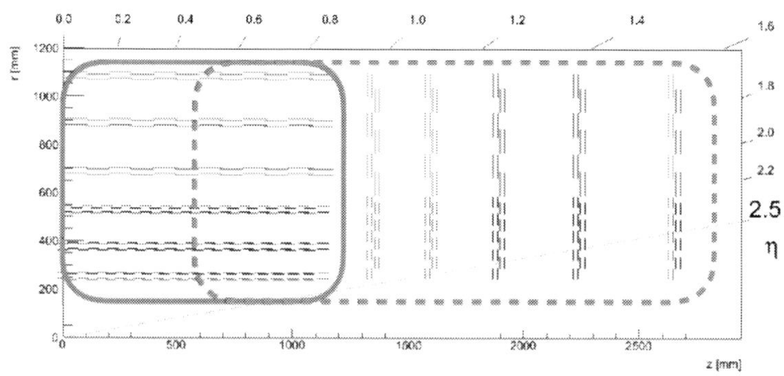

Figure 4. Sketch of the +z portion of the detector in the r–z plane. PS modules are shown in blue while 2S modules are shown in red. The box in blue shows the detector region covered by the half barrel implementation. A box in green shows the detector region covered by the other project that spans a quarter of the barrel, the transition region between the barrel and endcaps, and the endcaps.

A system hardware demonstrator was set up for full scale testing of the firmware implementation. The demonstrator was used to show that the full L1 tracking chain meets the required performance within the available latency. The demonstrator test stand is one slice of the $n = 6$ time multiplexed system. It includes three ϕ sector processing boards: one for the central ϕ sector, and one for each

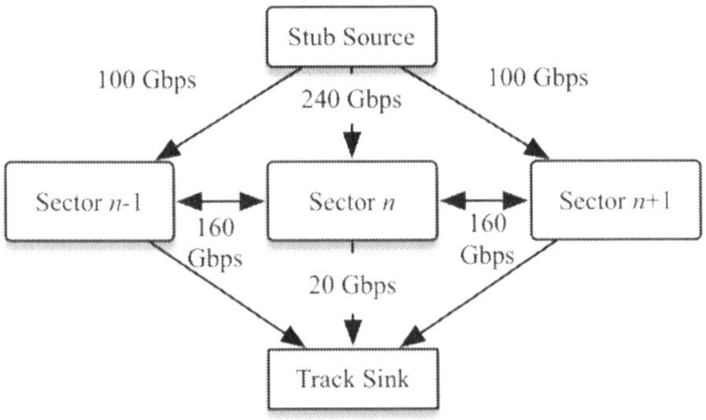

Figure 5. Schematic overview (left) and image (right) of the tracklet demonstrator system. The stub source and track sink are on the same processing board. Estimated data transfer rates between the boards are also shown.

of its nearest neighbors. One additional processing board has the duplicate function of sending the stubs into the ϕ sector boards and receiving the final track outputs. A schematic of the demonstrator is shown in Figure 5. The central ϕ sector processor is the actual system under test.

In the final system, each sector processor is foreseen to be an ATCA blade with a Virtex Ultrascale+ FPGA. The current demonstrator system instead is made of four μTCA blades, called CTP7 boards [8]. Each CTP7 has a Xilinx Virtex-7 (XC7VX690T) FPGA [9] and a Xilinx Zynq-7000 SoC processor for configuration and outside communication. The CTP7 boards were developed for the current CMS L1 trigger [10]. An AMC13 [11] card provides synchronization between the boards with a central clock distribution. The inter-board communication uses 8b/10b encoding with 10 Gbps links. The demonstrator system is shown in Figure 5.

3.3 Demonstrator System Latency

Each processing step outlined in Section 3.1 takes a fixed number of clock cycles to process its input data. Hence it is feasible to calculate a model of the latency for the complete system. Calculations are done assuming a 240 MHz clock and a time multiplexing factor of six, the current configuration of the project. The latency for each processing module to receive data and produce the first result varies between 1-50 clock cycles depending on the module. Each processing step continues to process data for the same event for 150 ns before switching to the next event.

For some of the steps, data has to be transmitted between boards. Tracklet projections and their corresponding matched stubs must be sent to the neighboring sector processors. In these cases, the latency due to inter-board communication and links is included in the latency model. The measured transmission latency is 316.7 ns (76 clock cycles). This latency includes all parts of the transmission: the transceiver TX and RX, channel bonding (the use of multiple serial links to send the data), data propagation through 15 m long optical fibers, and time needed to prepare and pass data from processing modules to transceivers. The latency also includes the data transmission latency for receiving stubs from and sending final tracks back to the data source/sink processing board.

A summary of the estimated latency is shown in Table 1. The total estimated latency for receiving the first track from an event is 3345.8 ns. Because of the fixed-processing time, the final processed track for any event will come within 150 ns of the first track. The total latency of the demonstrator has also been measured with a clock counter located on the data source/sink board. The measured first

Table 1. Demonstrator latency model. For each step, the processing time and latency is given. For steps involving data transfer, the link latency is given. The total model latency is 3345.8 ns.

Step	Proc. time (ns)	Step latency (CLK)	Step latency (ns)	Link delay (ns)	Step total (ns)
Input link	0.0	1	4.2	316.7	320.8
Layer Router	150.0	1	4.2	-	154.2
VM Router	150.0	4	16.7	-	166.7
Tracklet Engine	150.0	5	20.8	-	170.8
Tracklet Calculation	150.0	43	179.2	-	329.2
Projection Transceiver	150.0	13	54.2	316.7	520.8
Projection Router	150.0	5	20.8	-	170.8
Match Engine	150.0	6	25.0	-	175.0
Match Calculator	150.0	16	66.7	-	216.7
Match Transceiver	150.0	12	50.0	316.7	516.7
Track Fit	150.0	26	108.3	-	258.3
Duplicate Removal	0.0	6	25.0	-	25.0
Output Link	0.0	1	4.2	316.7	320.8
Total	1500.0	139	579.2	1266.7	**3345.8**

track out latency is 3333 ns, which agrees within three clock cycles (0.4%) with the modeled latency. This is already below the goal of 4 μs for L1 tracks.

There are notable places where the latency can be improved. First, the layer router has become redundant as the incoming stubs are now foreseen to be sorted by layer. Removing this processing module will reduce the latency by 150 ns. The transmission protocol contributes a large amount to the latency of the system. By sending duplicated stubs from the neighboring sectors to the central sector, the need for inter-board communication in the projection transceiver and match transceiver steps is removed. This could shave off as much as ≈ 1 μs from the total latency. Additional optimization in transmission protocol and clock speed will provide speedup as well.

4 Tracklet System Performance

The demonstrator showed excellent agreement between the actual firmware results and the integer-based C++ emulation of the system. For single object events, the output tracks from the firmware have 100% bitwise compatibility with the integer-based emulation. For busier events, for example top quark pair ($t\bar{t}$) events with an average PU = 200, the emulation and firmware tracks agree to better than 99%. Because of this, the C++ emulation can safely be assumed to emulate well the demonstrator system.

The estimated performance of the tracklet algorithm is studied with the integer-based emulation of the algorithm. The efficiency of reconstructing the input tracks with all of their correct stubs as a function of η for muon and electron events are shown in Figure 6. For all of these objects, efficiencies are shown for several different average pileup conditions, <PU> = 0, 140, and 200. The efficiency is computed as implemented in the demonstrator, i.e. it includes the effects of truncating data. Additionally the integer-based emulation provides an estimate of the efficiency without truncation effects. Efficiency with and without truncation effects are included in these plots. Here it is seen that for all of these objects the data truncation has little effect on the track finding efficiency. The effect is mini-

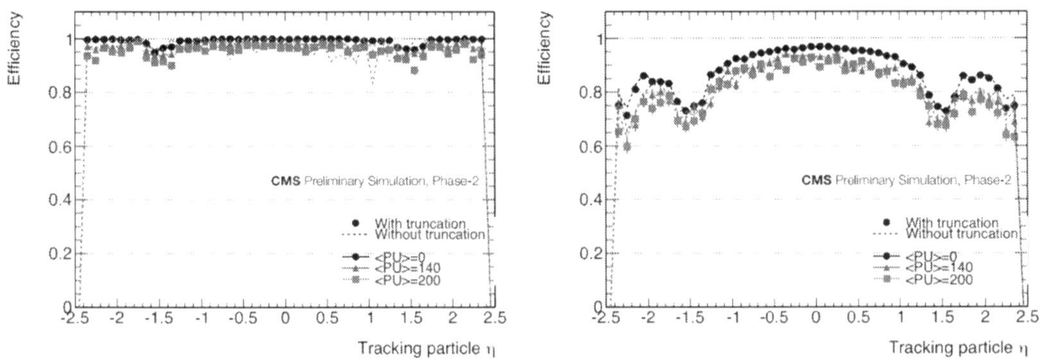

Figure 6. Efficiency as a function of η for muons (left) and electrons (right). Results are shown for three different average pileup scenarios: 0 (black), 140 (red), and 200 (blue). The efficiency is shown with (filled points) and without (dashed lines) truncation included.

mal for two main reasons: (i) because of the large parallelization of the system, most of the modules are sparsely populated, (ii) the different seeding combinations provide additional redundancy that can recover tracks that may otherwise be lost.

For very busy $t\bar{t}$ events there is a more drastic effect from truncation, as shown in Figure 7 (left). Each of the jets in these events are generally very dense (in ϕ) and therefore are not well-split between virtual modules. In the combinatoric heavy stages of the algorithm there is not enough time to process all of the stubs, causing a drop in the efficiency. This can be fixed by better load-balancing. Changing the partitioning of virtual modules so that they are thinner in ϕ but span the entire z of the detector alleviates these issues. With this new scheme, when there is a dense jet in an event, the stubs are spread over more virtual modules, meaning more stubs can be processed within the given latency and fewer stubs are lost due to truncation. The improvement in efficiency is shown in Figure 7 (right). Although this change increases by about 20% the number of virtual modules per ϕ sector, there is no additional resource usage. The partitioning is thin enough to completely remove one of the lookup tables in the calculation, and reduces the number of memories needed downstream for storing the projections. This improved load-balancing scheme is currently being implemented.

The tracklet algorithm achieves the resolution required of the L1 trigger system, specifically a z_0 resolution of 1-2 mm for $\eta < 1.9$ and a relative p_T resolution of less than 0.05 for almost the entire η range as shown in Figure 8. The integer-based emulation achieves comparable resolution to that of a floating-point simulation of the tracklet algorithm [6]. Some improvements can still be made. Too few η bins are used in the final track calculation lookup tables in the transition region between barrel and endcaps (the region with $\eta \approx 1.2$). This can be corrected, and will improve the resolution in that region. Additionally, the current version of the project uses too few bits for storing the stub z position. This can also be corrected and will improve overall the track z_0 resolution shown here.

While achieving high efficiency and good resolution, the tracklet approach also achieves a relatively low fake rate (the rate of incorrectly reconstructing a track). The total number of tracks found for single muon events with different average pileup scenarios is shown in Figure 9. Here it can be seen that for a single muon without pileup, the tracklet algorithm almost always finds only one track. The fake rate in this scenario is extremely low. With the addition of more pileup interactions, additional tracks are found. More detailed studies on rates and track isolation for the tracklet approach can be found in Ref. [3].

DOI: 10.1051/epjconf/201715000016

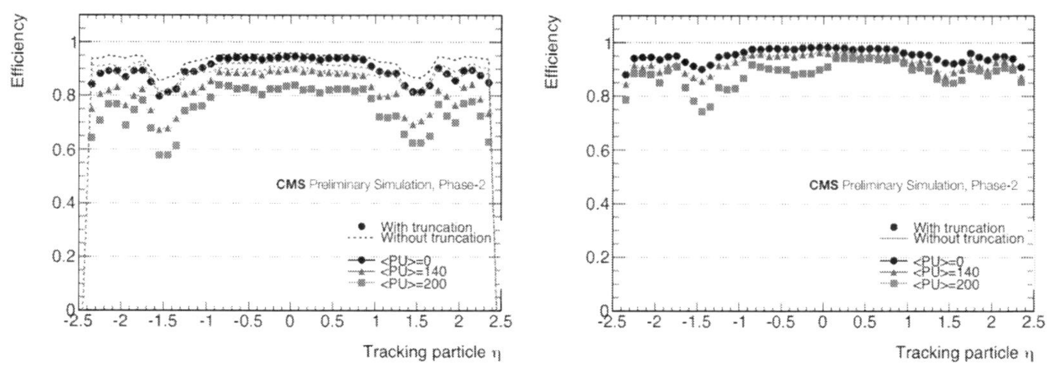

Figure 7. Efficiency as a function of η for $t\bar{t}$ events with standard configuration (left) and after improved load balancing (right). Results are shown for three different average pileup scenarios: 0 (black), 140 (red), and 200 (blue). The efficiency is shown with and without truncation included as points and dashed lines, respectively. The asymmetry between positive-negative η is introduced in the truncation and is currently under investigation.

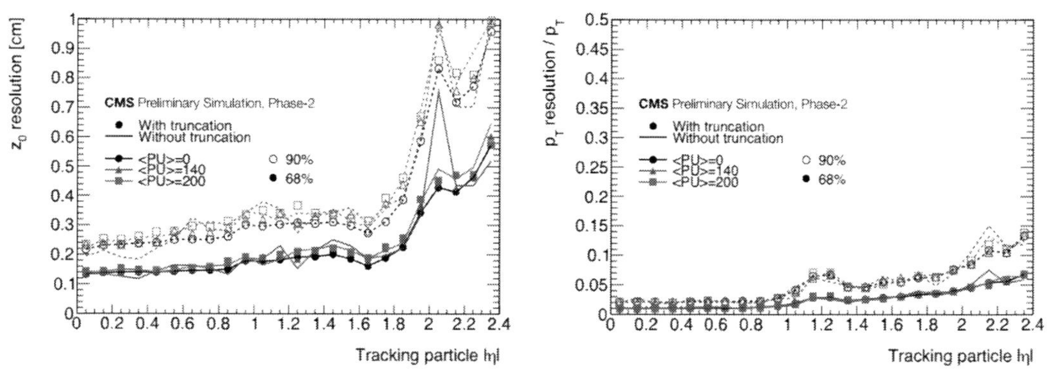

Figure 8. Resolution of z_0 (left) and relative p_T (right) for the final tracks for muon events. Results are shown for three different average pileup scenarios: 0 (black), 140 (red), and 200 (blue). The resolution is shown with and without truncation included as points and dashed lines, respectively.

5 Projections for a Full L1 Tracking System

Extrapolations can be made for a full tracklet track-trigger system to be used at CMS in the HL-LHC. Currently, almost an entire half ϕ sector (the $+z$ region of the detector) can be processed by a single Virtex-7 FPGA. Based on the resources available in this chip and what is used, it is anticipated that a full ϕ sector encompassing the full z range of the detector will be able to be processed by a single future-generation FPGA. Actually the Virtex-7 FPGA has enough lookup table (LUT) resources to store the precomputed portions of the algorithm for the full ϕ sector. Similarly this FPGA has the digital signal processing (DSP) resources needed to handle the relatively small number of calculations in the full algorithm. However, because the tracklet system relies largely on parallelization of the processing, each ϕ sector needs a lot of memory (in FPGAs these are fixed size block RAMs or BRAMs) to store intermediate steps in the calculation. The estimated resource usage for a full sector is shown in Table 2. As shown there, the needed resources, specifically the BRAM needs, surpass

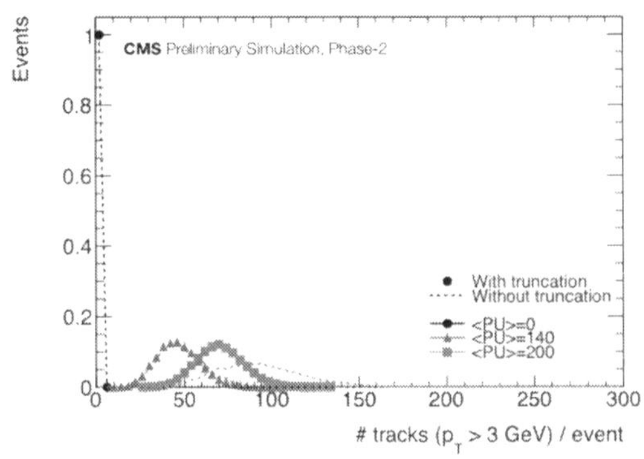

Figure 9. Total number of tracks (with $p_T > 3$ GeV) found for muons with three different average pileup scenarios: 0 (black), 140 (red), and 200 (blue). Results are shown with/without truncation included as filled points/dashed lines.

that of the current Virtex-7 FPGA. However, a full ϕ sector processing needs fit comfortably within the resource allotments of the Xilinx Virtex-Ultrascale+ class of FPGAs, a few of which are shown in Table 2. The project resource needs can also be met by a Kintex-Ultrascale (specifically KU115) FPGA, if 16Gbps (instead of 25Gbps) links are sufficient for sending stubs from the detector to the L1 tracking system.

Table 2. FPGA resource utilization for a full ϕ sector based on the usage as reported by Xilinx's Vivado Design Suite for the implementation of a half ϕ sector. The top line shows the needs for a full ϕ sector. The following lines show the fraction of the resource needs in Virtex-7 (top line) and Virtex Ultrascale+ (others) FPGAs.

	LUT Logic	LUT Memory	BRAM	DSP
Full sector	279733	151191	2721.5	1818
Virtex-7 690T	65%	87%	185%	51%
VU3P	32%	81%	85%	80%
VU5P	21%	53%	58%	52%
VU7P	16%	40%	42%	40%
VU9P	11%	27%	28%	27%
VU11P	10%	27%	29%	20%
VU13P	7%	20%	22%	15%

6 Conclusions

For the HL-LHC, CMS will require a new all-silicon tracker with trigger capabilities in order to achieve its physics goals. Tracking at the L1 trigger will provide greatly improved momentum measurements, enabling rate reducing triggers without increasing p_T thresholds. The tracklet approach is one of the proposed methods for performing track finding for the L1 trigger. The method is based on a road-search algorithm that is implemented on commercially available FPGA technology. The tracklet algorithm has been implemented as a floating-point simulation, integer-based emulation, and in an FPGA. Two projects are used to cover the entire $+z$ range of the detector. To demonstrate the system feasibility, a hardware demonstrator system based on Virtex-7 FPGAs was assembled and validated the algorithm's implementation. High performance, in terms of efficiency and resolution, is achieved. The demonstrator gives a measured latency for L1 track finding of 3.333 μs. With little extrapolation, the system seems scalable to future FPGA technology.

Acknowledgements

I would like to thank the CMS collaboration and contributors to this project at Cornell University, Rutgers University, Ohio State University, and University of Notre Dame. I also wish to thank the Wisconsin University CMS group for their support of the CTP7 platform. This work was supported by the US National Science Foundation through the grant NSF-PHY-1307256.

References

[1] F. Zimmerman, PoS **EPS-HEP2009**, 140 (2009)
[2] CMS Collaboration, JINST **3**, S08004 (2008)
[3] CMS Collaboration, Tech. Rep. CERN-LHCC-2015-010. LHCC-P-008. CMS-TDR-15-02, Geneva (2015), `https://cds.cern.ch/record/2020886`
[4] G. Hall, Nucl. Instr. Meth. A **824**, 292 (2016)
[5] G. Fedi, Eur. Phys. J. Conf. **127**, 7 (2016)
[6] L. Skinnari, EPJ Web Conf. **127**, 00017 (2016)
[7] J. Chaves, JINST **9**, C10038 (2014)
[8] A. Svetek et al., JINST **11**, C02011 (2016)
[9] Xilinx Incorporated, *7 Series FPGAs Overview*, http://www.xilinx.com/support/documentation/data_sheets/ds180_7Series_Overview.pdf, last accessed on 2017-04-09
[10] The CMS Collaboration, Tech. Rep. CERN-LHCC-2013-011 (2013)
[11] Boston University, *The AMC13 Project*, http://bucms.bu.edu/twiki/bin/view/BUCMSPublic/HcalDTC, last accessed on 2017-04-09

AUTHOR INDEX

AUTHOR INDEX

EDP Sciences – Web of Conferences
17, Avenue du Hoggar
Parc d'Activité de Courtabouf
BP 112
F-91944 Les Ulis Cedex A
France

ISBN 978-1-5108-4641-8

CURRAN ASSOCIATES INC.
proceedings.com

2nd International Conference on Language and Cultural Communication (ICLCC 2024)

SHS Web of Conferences Volume 185 (2024)

Sanya, China
19 – 21 January 2024

edp sciences